Environment and Sustainability

Celil Aydın & Burak Darıcı (eds.)

Environment and Sustainability

Bibliografische Information der Deutschen Nationalbibliothek
Die Deutsche Nationalbibliothek verzeichnet diese Publikation
in der Deutschen Nationalbibliografie; detaillierte bibliografische
Daten sind im Internet über http://dnb.d-nb.de abrufbar.

ISBN 978-3-631-77187-7 (Print)
E-ISBN 978-3-631-78370-2 (E-PDF)
E-ISBN 978-3-631-78371-9 (EPUB)
E-ISBN 978-3-631-78372-6 (MOBI)
DOI 10.3726/b15358

© Peter Lang GmbH
Internationaler Verlag der Wissenschaften
Berlin 2019
Alle Rechte vorbehalten.

Peter Lang – Berlin · Bern · Bruxelles · New York ·
Oxford · Warszawa · Wien

Das Werk einschließlich aller seiner Teile ist urheberrechtlich
geschützt. Jede Verwertung außerhalb der engen Grenzen des
Urheberrechtsgesetzes ist ohne Zustimmung des Verlages
unzulässig und strafbar. Das gilt insbesondere für
Vervielfältigungen, Übersetzungen, Mikroverfilmungen und die
Einspeicherung und Verarbeitung in elektronischen Systemen.

Diese Publikation wurde begutachtet.

www.peterlang.com

Preface

The concepts of sustainability, which constitute a new agenda in the discussions on economic development, have led to the questioning of the traditional perspective on environment and natural resources. Although it is possible to make different definitions on the concept of sustainability with many dimensions such as economy, energy and environment, the basic principle is to ensure continuity in these three different areas.

In this book, several specialist academicians in their field have examined the topic of "Environment and Sustainability" from different perspectives. The global environmental issues, environmental policies and the link to sustainability have been analyzed in both micro and macro scales. The current situation, changes and transformations from past to present are discussed both empirically and theoretically in the light of new approaches.

We would like to thank all the authors and researchers who contributed to the study for their commitment and scholarly nitpicking. We also appreciate the families and children who are the source of motivation for our authors with their patience and support during the conduct of these valuable works.

And also we would like to thank Professor Süleyman ÖZDEMİR, Rector of Bandırma Onyedi Eylül University for his contributions and support.

Sincerely yours

Editors; Celil AYDIN and Burak DARICI

Contents

List of Contributors 9

Hicran Özgüner Kılıç
The Effect of Environment on Marketing: Sustainable Marketing 11

Hatice Aydın and Derya Fatma Biçer
Anti-Consumption for Environmental Sustainability 25

Mutlu Uygun
The Green Brand Equity in terms of Environment and Sustainability 39

Halim Tatlı and Beşir Koç
The Relationship between Environmental and Socio-Demographic Factors that Affect Consumers' Demand for Goods 55

Mehtap Çakmak Barsbay and Aytuğ Altın
Environment and Healthcare Sector: Current Debates on Sustainability 75

Mustafa Cem Aldağ
Impacts of Technological Developments on the Environment and Agriculture 89

Metin Kılıç
An Outlook on Companies' Environmental Activities in terms of Corporate Governance 103

Mustafa Gül
Environmental Accounting 119

Meltem Ece Çokmutlu and Metin Kılıç
Evolution of Environmental Reporting: The Example of Turkey 133

Demet Beton Kalmaz
Economic Growth and Environmental Degradation in Turkey 149

Tunç Durmaz
Environmental Catastrophes and Deep Uncertainties Surrounding the
Economics of Climate Change 171

Ayşe Durgun Kaygısız
The Correlation between Environmental Pollution and Economic
Growth: Validity Analysis of the Environmental Kuznets Curve
According to the Panel Data Method 187

İlyas Okumuş and Abdulmecit Yıldırım
Investigating the Validity of the EKC Hypothesis in Eurasian
Countries: The Role of Financial Development 203

Celil Aydın, Burak Darıcı and Şeyma Şahin Kutlu
Economic Growth and Ecological Footprint: Reconsidering the
Empirical Basis of Environmental Kuznets Curves 221

List of Figures 241

List of Tables 243

List of Contributors

Mustafa Cem Aldağ
Dr., Bandırma Onyedi Eylül University, Balıkesir, Turkey. maldag@bandirma.edu.tr

Aytuğ Altın
Assistant Prof., Karamanoğlu Mehmetbey University, Karaman, Turkey. aytugaltin@gmail.com

Celil Aydın
Associate Prof., Bandırma Onyedi Eylül University, Balıkesir, Turkey. caydin@bandirma.edu.tr

Hatice Aydın
Associate Prof., Bandırma Onyedi Eylül University, Balıkesir, Turkey. haydin@bandirma.edu.tr

Derya Fatma Biçer
Assistant Prof., Cumhuriyet University, Sivas, Turkey. dfbicer@cumhuriyet.edu.tr

Mehtap Çakmak Barsbay
Assistant Prof., Karamanoğlu Mehmetbey University, Karaman, Turkey. mehtapcakmak@kmu.edu.tr

Burak Darıcı
Prof., Bandırma Onyedi Eylül University, Balıkesir, Turkey. bdarici@bandirma.edu.tr

Ayşe Durgun Kaygısız
Assistant Prof., Süleyman Demirel University, Isparta, Turkey. aysedurgun@sdu.edu.tr

Tunç Durmaz
Assistant Prof., Yıldız Technical University, İstanbul, Turkey. tdurmaz@yildiz.edu.tr

Meltem Ece Çokmutlu
Research Assistant, Karabük University, Karabük, Turkey. meltemece@karabuk.edu.tr

Mustafa Gül
Asisstant Prof., Gaziosmanpaşa University, Tokat, Turkey. mustafa.gul@gop.edu.tr

Demet Beton Kalmaz
Assistant Prof., European University of Lefke, Lefke, Northern Cyprus. demetkalmaz@eul.edu.tr

Hicran Özgüner Kılıç
Assistant Prof., Bandırma Onyedi Eylül University, Balıkesir, Turkey. hkilic@bandirma.edu.tr

Metin Kılıç
Assistant Prof., Bandırma Onyedi Eylül University, Balıkesir, Turkey. metinkilic@bandirma.edu.tr

Beşir Koç
Assistant Prof., Bingöl University, Bingöl, Turkey. bkoc@bingol.edu.tr

İlyas Okumuş
Assistant Prof., Mustafa Kemal University, Hatay, Turkey. ilyasokumus@mku.edu.tr

Şeyma Şahin Kutlu
Research Assistant, Bandırma Onyedi Eylül University, Balıkesir, Turkey. ssahin@bandirma.edu.tr

Halim Tatlı
Assistant Prof., Bingöl University, Bingöl, Turkey. htatli@bingol.edu.tr

Mutlu Uygun
Assistant Prof., Aksaray University, Aksaray, Turkey. mutluuygun@gmail.com

Abdulmecit Yıldırım
Assistant Prof., Muş Alparslan University, Muş, Turkey. a.yildirim@alparslan.edu.tr

Hicran Özgüner Kılıç

The Effect of Environment on Marketing: Sustainable Marketing

Introduction

Enterprises, by their nature, both affect their environment and are affected by it. This is a requirement of being an open system. Marketing success of enterprises depends mostly on the ability of managers to manage marketing plan and programs in this environment. As far as the interaction between marketing and the environment is taken into consideration, it is necessary to continuously observe and evaluate the developments and changes in the environment. When the environmental factors are evaluated with regard to natural resources and pollution, it is seen that the balance of life changes negatively each passing day. The efforts of turning the balance into a positive direction, namely nature and environment consciousness, started especially during the 1980s with the enforcement of the law on environment protection and continued during the 1990s with the integration of environment-friendly product and services into the market and has developed so far today. Protection of environment and environment consciousness can be explained with the notion of sustainability. Sustainability can be defined as the protection process of some factors required by social, economic and ecologic systems. Sustainable marketing that emerges as an application of relationship marketing approach, the last step of the development of marketing, is a structure that supports this process. Sustainable marketing can be defined as establishing and sustaining sustainable relations with consumers, social environment and natural environment. The sensitive enterprises that give importance to the protection of the environment and consumers should behave accordingly in all marketing activities such as developing, pricing, distributing and promotion of products and services.

In this respect, this study is going to deal with the relationship between marketing and the environment, environmental change in the structure of the market, development process of a sustainable marketing, the notion of sustainable marketing and sustainable marketing mix.

1 Relationship between Marketing and Environment

In order to put forth the relationship between marketing and environment, it is necessary to talk about two main marketing approaches. These approaches are traditional marketing and relationship marketing. Traditional marketing is defined as "the process of planning and applying the efforts of creating, pricing, distributing and selling products, services and ideas in a way that will ensure change in accordance with the aims of persons and organizations." To put it differently, it is defined as the sum of organization activities that direct goods and services from manufacturers to consumers or users for fulfilling the goals of enterprises and satisfying the needs and desires of current customers and potential consumers (Dinçer, 2012: 1). Since the 1970s when the approach of traditional marketing used to be dominant, serious arguments have been made about the relationship between marketing and the environment. Marketing has two roles, namely the role in stimulating the demand and consumption on the unsustainable level and the role in activating market mechanisms to tackle social and environmental problems (Peattie, 2001: 129). These two roles contradict with each other. Enterprisers attempt to both encourage consumption and solve the problems resulting from over consumption. It should also be mentioned that traditional marketing approach has a significant role in the development of these ecologic problems. The reason for this is considered to result from the facts that the traditional marketing approach causes over consumption of products and services, that the system ignores environmental factors, that status is believed to increase with the assets possessed, that short-term profit maximization is aimed and a profit-oriented approach is adopted and that the life course of product is quite short because of resource-waste cycle (Pezikoğlu, 2010: 827).

According to the relationship marketing approach, marketing is defined as a social science that aims to form and sustain long-term relations that add value to its stakeholders and establish communication and interaction (Erdoğan, 2014: 5). Departing from these definitions, the enterprises, within the plan and programs of their marketing activities, need to develop and offer products and services that look out for the desire and needs of all their stakeholders and be careful to use the available resources and opportunities optimally during that process. Especially during the 1980s, the enterprises started to find out some ways to use available resources effectively and please the stakeholders. The number of international non-profit organizations increased, governments carried out some activities and a public awareness started to develop (Aytekin, 2007: 1). All these pressures and efforts, limitless desire and needs of consumers and the scarcity of resources increased the importance of environment consciousness during the

1990s. The enterprises aim to provide information to the consumers on environment and nature friendly products, behave with the consciousness of using the available limited resources in an effective manner, direct the activities of the enterprises and fulfill the desires of consumers (Ayyıldız and Genç, 2008: 506). These efforts contribute to the development and maintenance of long-term relations with the stakeholders. The sustainable marketing that emerges as an application of relationship marketing, the last step of marketing development, is a structure that supports this process.

Enterprises are effective in directing marketing activities and the behaviors of their target group. For this reason, they fulfill a significant function in increasing the marketing life standards and creating environment consciousness around the world (Onaran, 2014: 42). Integrating sustainable marketing practices into the approach of relationship marketing, creating environment consciousness and giving importance to the approach of nature protection have prompted the people or organizations interested in these issues and led to the integration of the environment related criteria into international quality standards and legal regulations (Armağan and Karatürk, 2014: 3). These organizations, standards and regulations adopt environmental consciousness in all of the processes including product supply and usage and the application of recycling activities. They aim to provide environmental benefits in all products and production activities that affect all living creatures and wildlife, exist and will exist in the market (Özçelik, 2017: 5). As a result, compliance with the relevant regulations on the protection of environment turns into an obligation and shows increase. Some of these organizations are the United Nations Development Program (UNDP), the Food and Agriculture Organization, NATO, the Organization for Economic Cooperation and Development (OECD), Council of Europe, the International Trade Organization, IMF and the World Bank. Stockholm United Nations Conference on Human and Environment, Brundtland Report, Rio United Nations Conference on Environment and Development, Johannesburg World Summit of Sustainable Development and Kyoto Protocol can be given as examples for some of the global efforts (Yılmaz and Güney, 2015: 239).

2 Environmental Factors in the Birth and Development Process of Sustainability

The notion of sustainability emerged first in the document of World Charter for Nature accepted by the International Union for the Conservation of Nature (IUCN) in 1982. According to this charter, it is foreseen that the ecosystem, organisms, land, sea and atmosphere resources should be managed in a way to obtain optimum sustainability without putting the integrity of ecosystems and species into danger (Mısırdalı Yangil, 2015: 358).

The current definition of the notion of sustainability was provided in 1987 in Brundtland Report prepared by the World Commission on Environment and Development. The issue that is emphasized in the notion of sustainability about the environment is to enable and manage the usage of natural resources optimally without putting the lives of living beings into danger. In the formation of sustainability, various environment-related economic and social problems such as increase in worldwide pollution and population, resource depletion, global warming as well as increase in the need for energy generating resources are observed and (Adanacıoğlu, 2015: 596) solutions for them are tried to be found. It came out as ecologic marketing in 1975 during the first period of sustainable marketing development process under the auspices of American Marketing Union (Çakır, 2017: 336). Ecologic marketing can be defined as marketing activities aiming to prevent such environmental problems as natural resources depletion and air pollution with some technological and social opportunities (Armağan and Karatürk, 2014: 3). During those years, the notion of ecological marketing was mostly related to decreasing specific environmental problems in the industries that have the biggest environmental effect by developing new technologies (Peattie and Charter, 2003: 727). The second period of the 1980s was related to environmental marketing that focuses on defending clean energy, understanding and targeting green consumers, displaying a good socio-environmental performance as a potential of competitive advantage and encouraging marketers to obtain the physical system appearance of enterprises (Hunt, 2011: 7). Environmental marketing denotes marketing activities that enterprises carry out by paying more attention to the natural environment (Çabuk and Nakıpoğlu, 2003: 42). Environmental marketing has enlarged its scope by including cleaning products, white goods, carpet, paper etc., and such services as banking and tourism (Peattie, 2001: 134). The environmental disasters that have been experienced since the 1970s and 1980s have continuously increased the interest in the environment (www.ntv.com.tr). During the same years with those disasters, Article 56 of the 1982 Constitution of Turkey included the following provision: "Every person has the right to live in a healthy and balanced environment and improving the environment, protecting its health and preventing environmental pollution is the duty of the state and citizen". In addition, the Environmental Law dated 1983 and numbered 2872 (www.cevreonline.com), deals with the issue of environment with an approach accepted in many developed countries by charging the state and individuals with the duty of protecting and improving the environment. Apart from these, "the right of living in a healthy environment" accepted

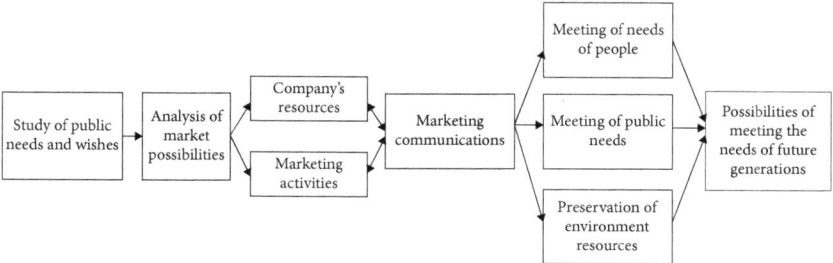

Fig. 1: The Notion of Sustainable Marketing. Source: Praude and Bormane, 2014: 168.

by the organization of international consumers union was announced worldwide as one of the international rights of the consumers (Atalay Oral and Akpınar, 2011: 3). These legal amendments gained a larger dimension with the experienced disasters, the efforts to slowly develop an environment consciousness, discussions on environment that emerged at the beginning of the 1990s, extinctions of species, destruction of some ecosystems as well as poverty and famine (Ar Akdeniz, 2011: 57). This enlarged dimension of the third period refers to the sustainable marketing, which is the marketing dimension of the notion of sustainability foregrounded in many areas such as economy and finance.

3 The Notion of Sustainable Marketing

Sustainable development is defined as a kind of development that "meets the needs of today without depriving the future generations from their ability to meet their own needs". For this reason, sustainable marketing is defined as "marketing activities that support a sustainable economic development" (Hunt, 2011: 7).

Various definitions have been provided for the definition of sustainable marketing. G. Armstrong and P. Kotler claim that sustainable marketing "is socially responsible and environment-friendly marketing that meets the current needs of consumers and businesses while preserving or even improving the ability of future generations to meet their future needs" (Krukowska-Miler, 2017: 2). The notion of sustainable marketing is mentioned in Fig. 1.

In Fig. 1, the notion of sustainability is based on the principle that the fulfillment of current needs and wishes of consumers depends on market conditions, company's resources, marketing activities, communication with consumers, foreseeing the protection of environmental resources and their development (Praude and Bormane, 2014: 168). In other words, the main principle behind sustainable marketing is to put forth the decisions with which environmental problems are emphasized and environment forces are used in a competitive manner. The notion of sustainable marketing aims to meet the needs of future generations, which means creating a sustainability-based value for future generations, forming and distributing communication (Capatina, Micu, Cristache and Micu, 2017: 288).

Sustainable marketing can play a significant role in the development of positive aspects of marketing, decreasing negative ones and contributing to its long-term success. (Hurth and Whittlesea, 2017: 360).

The mission and vision of a company needs to support sustainable marketing that exists within its main values. It is not only under the responsibility of marketing department to design, produce and offer a sustainable product, but it also requires the collaboration of all departments in a company. For the company to obtain real sustainability, it is expected to establish coordination among all members of the supply chain in accordance with the consumers' expectations (Bridges and Wiihelm, 2008: 35).

4 Sustainable Marketing Mix

While the marketing mix had been associated with product-oriented 4P since the existence of traditional marketing understanding, it gave place to 4C, the customer-oriented marketing mix. With this change, customer value was expressed as price customer cost and promotion customer communication while distribution was expressed as customer convenience in the literature (Kotler, 2005: 120). 4C, which was mentioned in the formation of this new marketing mix, considers environmental and social aspects of the purchased product with a customer-oriented approach (Reutlinger, 2012: 27). The sustainable marketing mix that was developed in order to form and apply marketing strategies is explained in Tab. 1 with the notions of customer value, customer cost, customer communication and costumer convenience (Özbakır, 2010: 59).

Tab. 1: Sustainable Marketing Mix Elements. Source: Ozbakir, 2010: 59.

Customer Value	Presenting products and services that bears value to customers
Customer Cost	Taking into account the cost to environment and customer of products and services
Customer Communication	Providing the transmission of the values of products and services by honest, lucid and open businesses
Customer convenience	Developing optimum conditions to customers in their access and usage of products and services.

4.1 Customer Value

Customer value is the change value realized between the benefits the products provide to the customers and the compromises the customers endure in order to obtain it (Çalhan, Çakıcı and Karamustafa, 2012: 90). As the benefits of the purchased products to the customer increases, its value will increase proportionally. As a result, it is important to establish trust-based relations with customers, meet their needs and wishes in a sustainable manner and offer valuable products and services. Developing sustainable products and services is explained with reference to such topics as creating primary demand, determining main product functions, enabling the function of the product during its life cycle, being a friend to the nature, selecting appropriate resources, using and distributing them in an effective way etc. (Onaran, 2014: 96).

In production activities of enterprises, environmentally hazardous waste material accompanies the product as a result of the used material, energy, raw material and technology usage. The sensitive enterprises that make an effort for the minimum damage and recycling of this waste material can offer ecological products within the scope of sustainable marketing (Çelik, Akman, Ceyhan and Akman, 2016: 280).

Sustainable products need to contain some features in themselves. The main features of a sustainable product include "preventing pollution, protecting environment and designing industrial products" (Onaran, 2014: 97–99):

Enterprises can contribute to sustainability by designing environment-friendly products during the processes of product development and production and by managing all of their environmental responsibility activities in an integrated manner (Zeren and Nakıpoğlu, 2009: 460). The number of such enterprises has been increasing day by day. It is possible to exemplify new generation E.C.A brand ECOLOGIC armatures, its special design toilet made from antibacterial material (www.eleks.com.tr), the production of Pet Tambur, which came into existence with Grunding's environment-friendly design, with recycled plastic bottles and washing machines of Vestel Pyrojet technology A+++ energy class (www.capital. com.tr) as environmentalist approaches in sustainable product design and production processes.

Another element that creates customer value is packaging. Packaging has the function of protecting the product from external factors, decreasing the harm that it will get and easing its storage and distribution, enabling one to get informed of the brand and personalizing it. However, it can also be interpreted as a factor that leads to piles of rubbish in the environment and contaminates water, land and briefly the nature. For this reason, it is necessary to develop the packaging features with a sustainable marketing perspective (Onaran, 2014: 11–12). In order to develop environment consciousness in consumers, such enterprises as Carrefour, BİM and Migros need to serve as a model for sustainable packaging by using cloth bags.

The label that is placed on the package or exists independently is a source of information for consumers about the product's usage purpose, shape, brand weight, amount, features, production and expiration date. Environment sensitive version of labeling is called ecolabeling. Ecolabeling aims to inform consumers of the products that are less environmentally hazardous. In addition, International Standardization Organization (ISO) introduced 14000 documents series so that environment-friendly products could be differentiated and gain a specific status. The products that deserve these documents are accepted by the consumers as environment-friendly products (Yücel and Ekmekçiler, 2008: 329).

Ecolabeling is dealt in three aspects as ecolabelling, disposable labels and negative effects (Başaran Alagöz, 2007: 6–7). A successful ecolabeling protects the environment because of the fact that labeled products are more sensitive to the environment than the others and encourage consumers to decrease their usage of environmentally hazardous products and purchase environment-friendly ones.

4.2 Customer Cost

In light of the worldwide environmental changes, individuals have understood that living in a healthier environment by minimizing the negative factors requires an extra cost and that they can save themselves from this irremediable outcome by purchasing environment-friendly products. The thought that consumers are willing to pay more for environment-friendly green products has also increased the customer cost of the product (Kozanlıoğlu, 2010: 13).

Before adopting a sustainable price policy, the enterprises need to pay attention to such issues as "quality, being reliable, simplicity, being marketable, specificity, visibility, collectivism, strategy and resolution" (Uydacı, 2011: 214).

Total customer cost is the sum of time, money and energy. Purchase cost consists of investigating about the product and evaluating alternative ones as well as transportation cost. Purchase cost shows difference according to the amount of information about the sustainable product, the easiness of comparing alternative products and their accessibility. When the product is started to be used, energy, maintenance-repair costs, differences among similar products etc. can affect consumers' preferences. When the product's lifetime ends, disposition or shipment cost of the product comprises the costs after usage (Reutlinger, 2012: 31).

The price that costumers pay during purchasing a product or service may not always meet the cost of this product. The products and serviced offered to the consumers sometimes can be sold at a price less than its cost. However, social or ecologic environment may endure these cost whose effects are not realized by the consumers. Many social and environmental costs such as child or uninsured workers employed during the production process in order to decrease costs, factory's environmentally hazardous wastes, transportation costs for imported raw materials and air pollution are not reflected in prices. However, social and environmental costs and the cost that occurs at the beginning and end of the product's lifetime need to be included in the total customer costs (Özbakır, 2010: 69–70).

4.3 Customer Communication

Customer communication can be defined as communication arrangements that give value to customers within the scope of such promotion activities as advertisement, public relations, personal selling, sales development and direct marketing.

The customer-oriented enterprises create databases in order to understand the needs and wishes of customers better, develop solutions for their problems simultaneously and maintain their communication. In these databases, such information as age, gender, occupation, education of customers, the frequency of their shopping, the products they prefer, the time and place of their shopping, contact and address information etc. are stored. According to the obtained information, all contact channels related to the personalized product and services are directed. As a result, a continuous communication is established with the customer by offering the products and services that they prefer whenever and wherever they want (Aslan Çetin, 2018: 450).

In advertisement contents prepared for the communication activities with consumers, public relations campaigns and sales promotion activities as well as in all of the issues of product features, packages and distribution, it is significant to give the message that protection of the environment is regarded and environment-friendly attitudes are adopted, and to create consistency with all of the communication channels and tools for a sustainable communication (Erbaşlar, 2012: 98).

4.4 Customer Convenience

The availability of product and services at points to which customers can easily reach is possible with the effective sustainment of distribution activities.

Otherwise, in case of a problem in the distribution line, it is highly improbable for the product to stay in the market for a long time no matter how strong, functional and qualified the products developed by the enterprises are. On the other hand, if it is aimed to offer a sustainable product to the market, more attention needs to be paid to the issue of distribution.

The distance between the producer and consumer, a transportation infrastructure that functions as a bridge and the mediators' storage methods are significant factors in terms of their impact of physical distribution on the environment and customer (Özbakır, 2010: 71). At this point, with the development of technology, customers can easily reach information on the products they demand through such tools as websites, e-mail and social media. In addition, with the supply of data, the burden on physical distribution will be decreased and some contribution to the sustainable distribution will be made by saving on resource usage (Demirel, 2017: 112).

Minimizing the costs resulting from distribution activities, as well as the environmentally hazardous effects that may occur during product supply and transportation process, and considering this effect in the tools to be used, affect the status of enterprises in the market (Onurlubaş and Dinçer, 2016: 55). Paying attention both to the future and now and being sensitive to the society, the enterprises leave a positive impression on consumers, which is realized in their distribution policies. For instance, fuel saving in product distribution, minimizing product volume with innovative methods, enabling customers to reach sales points by benefiting from technological innovations and opportunities etc. can be evaluated in this context (Türk and Gök, 2010: 207).

In achieving their target, the enterprises need to consider their stakeholders in the supply chain that involves all kinds of activities starting from the production of products and services to the ultimate consumer. It results from the fact that half of what brings out the value of product consist of the suppliers. Sustainability of supply chain is defined as the management and encouragement of environmental, social and economic effects during the lifetime of product and services. Ten principles were determined in the United Nations Global Compact, which was prepared to enable this sustainability. It is also considered a ground to encourage the enterprises for sustainability (Onaran, 2014: 120–121).

In sustainable distribution activities, there exists a flow not only from producer to consumer, but also from consumer to producer. This flow is called reverse logistics. Reverse logistics is the systematic acceptance process of the products or items that are sent from a consumption point for potential recycling, reproduction or disposition (Karaçay, 2005: 318). This system operates by activating expired products and worthless resources that have not been used before or do not provide any benefit any longer by means of the recycling or reproduction methods. As a result, wastes are utilized and available resources are used economically (Fettahlıoğlu and Birin, 2016: 92). With reverse logistics activities, enterprises can both save on their costs and give less harm to the environment. By using the resources in an effective way, the enterprises contribute to sustainable development, gain the opportunity for creating an environment-friendly positive image and gain a significant place in the application of sustainable distribution. The sectors where reverse logistics is used include Automotive, Steel, Electronic, Computer, Chemistry, Pharmacy, Tire and Medical Sector. Many companies have carried out some works in terms of sustainable distribution activities (Doğanay and Kırcova, 2015: 409).

Conclusion

Human beings interact with the environment in which they live. Humans' harm to the balance in their living space has been the starting point of environmental degradation and problems. As the balance gets harm, many problems such as air, water and land pollution, quick depletion of natural resources, global warming etc. occur. While preparing the plan and programs of marketing activities, the enterprises need to evaluate environmental changes. At this point, the society needs to become conscious and the enterprises need to develop activities that are sensitive to environmental changes.

Adopting sustainable understanding has a significant role in the protection of environment and awareness increase. The sustainable marketing, one of the applications of relationship marketing, is a structure that supports the process of forming and maintaining relations with social environment and natural environment. The enterprises adopting such understanding need to behave sensitively by protecting the environment and consumes with the products and services they offer to the market.

In conclusion, it is necessary for both general public and enterprises to use their resources in an effective and economic way. While meeting the needs and wishes of customers, the enterprisers should behave with environment consciousness and develop appropriate policies and strategies. In this way, some contribution will be provided for a sustainable future.

References

Adanacıoğlu, H. (2015) "Sürdürülebilir Tarımsal Pazarlama Girişimleri", Türk Tarım-Gıda Bilim ve Teknoloji Dergisi, 3 (7): 595–603.

Alçın, S. (2016) "Üretim İçin Yeni İzlek: Sanayi 4.0", Journal of Life Economics, (3):2 19–30.

Ar Akdeniz, A. (2011) "Yeşil Pazarlama", Beta Yayınları, İstanbul.

Armağan, E. ve Karatürk, H.E. (2014) "Yeşil Pazarlama Faaliyetleri Çerçevesinde Aydın Bölgesindeki Tüketicilerin Çevreye Duyarlı Ürünleri Kullanma Eğilimlerini Belirlemeye Yönelik Bir Araştırma", Organizasyon ve Yönetim Bilimleri Dergisi, 6 (1): 1–17.

Aslan Çetin, F. (2018) "Sürdürülebilir Pazarlama: Tüketicilerin Sürdürülebilir Tüketime Yönelik Satın Alma Davranışları", Social Sciences Studies Journal, 4 (14): 447–455.

Atalay Oral, M. ve Akpınar, M.G. (2011) "Sürdürülebilir Pazarlamada Tüketim Kültürü: Tarım Ürünleri Örneği", Dünya Gıda E-Dergi, http://www.dunyagida.com.tr/haber/surdurule bilir-pazarlamada-tuketim-kulturu-tarim-urunleri/3955, Access Date: 23.05.2018.

Aytekin, P. (2007) "Yeşil Pazarlama Stratejileri", Celal Bayar Üniversitesi Sosyal Bilimler Enstitüsü Dergisi, 5 (2): 1–20.

Ayyıldız, H. ve Genç, K.Y. (2008) "Çevreye Duyarlı Pazarlama: Üniversite Öğrencilerinin Çevreye Duyarlı Pazarlama Uygulamaları ile İlgili Tutum ve Davranışları Üzerine Bir Araştırma", Atatürk Üniversitesi Sosyal Bilimler Enstitüsü Dergisi, 12 (2): 505–527.

Başaran Alagöz, S. (2007) "Yeşil Pazarlama ve Eko Etiketleme", Akademik Bakış, Uluslararası Hakemli Sosyal Bilimler E-Dergisi, 11, 1–13, https://arastirmax.com/en/ bilimsel-yayin/ akademik-bakis/11/ 1-13-yesil-pazarlama-eko-etiketleme.

Bridges, C.M. and Wilhelm, W.B. (2008) "Going beyond Green: The "Why and How" of Integrating Sustainability into the Marketing Curriculum", Journal of Marketing Education, 30 (33): 33–46.

Capatina, A., Micu, A., Cristache, N. and Micu, A.E. (2017) "The Impact of a Trend Pattern for Sustainable Marketing Budgets on Turnover Dynamics", Contemporary Economics, 11 (3): 287–302.

Çabuk, S. ve Nakipoğlu, M.A.B. (2003) "Çevreci Pazarlama ve Tüketicilerin Çevreci Tutumlarının Satın Alma Davranışlarına Etkileri ile İlgili Bir Uygulama", Çukurova Üniversitesi Sosyal Bilimler Dergisi, 12 (12): 39–54.

Çakır, M. (2017) "Yeşil Ürün Grupları Çerçevesinde Marka ve Markalama Kararları", Siirt Üniversitesi Sosyal Bilimler Enstitüsü Dergisi, 9: 333–378.

Çalhan, H., Çakıcı, A.C. ve Karamustafa, K. (2012) "Müşteri Değeri, Müşteri Sermayesi ve Otel Performansı İlişkisi Üzerine Kavramsal Bir Değerlendirme", Erciyes Üniversitesi Sosyal Bilimler Enstitüsü Dergisi, 1 (33): 87–120.

Çelik, İ.E., Akman, Ö., Ceyhan, A. ve Akman, V. (2016) "Yeşil Pazarlamada Sürdürülebilirlik ve Dünya'dan Bir Örnek: Tchibo", International Conference on Eurasian Economies, Session 4B: 278–282.

Demirel, Y. (2017) "Müşteri İlişkileri Yönetimi", Seçkin Yayınevi, Ankara.

Dinçer, Ö. (2012). "Pazarlama Nedir?", http://www.pazarlamamakaleleri.com/pazarlama-nedir/, Access Date: 10.06.2018.

Doğanay, Ö. ve Kırcova, İ. (2015) "Daha Yaşanılabilir Bir Dünya için Sürdürülebilirlik", 20.Ulusal Pazarlama Kongresi, Bildiri Kitabı, 10–13 Haziran, Eskişehir, 403–412.

Erbaşlar, G. (2012). "Yeşil Pazarlama", Mesleki Bilimler Dergisi, 1 (2): 94–101.

Erdoğan, B.Z. (2014) "Pazarlama", Ekin Yayınevi, Bursa.

Fettahlıoğlu, H.S. ve Birin, C. (2016) "Sürdürülebilirlik Açısından Tersine Lojistik Faaliyetleri ve Sürdürülebilir Pazarlamayı Etkileyen Faktörlerin Analitik Hiyerarşi Yöntemi ile Belirlenmesi", Kahramanmaraş Sütçü İmam Üniversitesi İktisadi ve İdari Bilimler Fakültesi Dergisi, 6 (2): 89–114.

Hunt, S.D. (2011) "Sustainable Marketing Equity and Economic Growth: A Resource- Advantege, Economic Freedom Approach", Journal of the Academy of Marketing Science, 39 (1): 7–20.

Hurth, V. and Whittlesea, E. (2017) "Characterising Marketing Paradigms for Sustainable Marketing Management", Social Business, 7 (3–4): 359–390.

Karaçay, G. (2005) "Tersine Lojistik: Kavram ve İşleyiş", Çukurova Üniversitesi Sosyal Bilimler Enstitüsü Dergisi, 14 (1): 317–331.

Kotler, P. (2005). "A'dan Z'ye Pazarlama", Çeviren: Aslı Kalem Bakkal, MediaCat Yayınları, İstanbul.

Kozanlıoğlu, B. (2010) "Sürdürülebilir Temelli Kurumsal Sosyal Sorumluluk Projelerinin Tüketici Tercihleri Üzerine Etkisi", Dumlupınar Üniversitesi, Sosyal Bilimler Enstitüsü, İşletme Anabilim Dalı, Yayınlanmamış Yüksek Lisans Tezi, Kütahya.

Krukowska-Miler, A. (2017) "Sustainable Marketing in the Health Care Organizations" Economic Processes Management", 2, 1–9, http://epm.fem.sumdu.ua/download/2017_2/epm2017_2 _ 11.pdf.

Mısırdalı Yangil, F. (2015) "Kurumsal Sürdürülebilirlik Kapsamında Sürdürülebilirlik Raporlarına Yönelik İçerik Analizi: Türkiye'deki En Büyük 100 Sanayi İşletmesi", İşletme Araştırmaları Dergisi, 7 (3): 356–376.

Onaran, B. (2014) "Sürdürülebilir Pazarlama", Detay Yayıncılık, Ankara.

Onurlubaş, E. ve Dinçer, D. (2016). "Yeşil Pazarlama", Beta Yayınları, İstanbul.

Özbakır, M. (2010) "Sürdürülebilir Pazarlamaya Geçiş ve Bu Sürece Yönelik Bir Uygulama: Marks & Spencer Örneği", Abant İzzet Baysal Üniversitesi, Sosyal

Bilimler Enstitüsü, İşletme Anabilim Dalı, Yayınlanmamış Yüksek Lisans Tezi, Bolu.

Özçelik, S. (2017) "Sürdürülebilirlik ve Yeşil Pazarlama", İstanbul Ticaret Üniversitesi, Dış Ticaret Enstitüsü, WPS No: 100-5.

Peattie, K. (2001) "Towards Sustainability: The Third Age of Green Marketing", The Marketing Review, 2 (2): 129-146.

Peattie, K. and Charter, M. (2003) "Green Marketing, The Marketing Book", (Edited by: Micheal J. Baker)", Butterworth-Heinemann, Great Britain, UK, 726-755.

Pezikoğlu, F. (2010) "Sürdürülebilir Kalkınma ve Yeşil Pazarlama", Türkiye IX. Tarım Ekonomisi Kongresi, Bildiri Kitabı, Şanlıurfa, 824-831.

Praude, V. and Bormane, S. (2014) "Sustaınable Marketing – Prospects and Challenges under Present Economy" Regional Formation and Development Studies, 3 (11): 165-176.

Reutlinger, J. (2012) "Sustainable Marketing: The Importance of Being a Sustainable Business" Lahti University of Applied Sciences Degree Programme in International Business, Bachelor's Thesis.Türk, M. ve Gök, A. (2010) "Yeşil Pazarlama Anlayışı Açısından Üretici İşletmelerin Sosyal Sorumluluğu", Elektronik Sosyal Bilimler Dergisi, 9 (32): 199-220.

Uydacı, M. (2011) "Yeşil Pazarlama" Türkmen Kitabevi, İstanbul.

Yılmaz, O. Y. ve Güney, C. (2015) "Yeşil İşletme Fonksiyonlarından Yeşil Pazarlama ve Yeşil Muhasebe" 3. Bölgesel Kalkınma Konferansı Bildiri Kitabı, Bingöl, 233-248.

Yücel, M. ve Ekmekçiler, Ü.S. (2008) "Çevre Dostu Ürün Kavramına Bütünsel Yaklaşım; Temiz Üretim Sistemi, Eko-Etiket, Yeşil Pazarlama", Elektronik Sosyal Bilimler Dergisi, 7 (26): 320-333.

Zeren, D. ve Nakıpoğlu, G. (2009) "Sürdürülebilir Ürün Tasarımında Tanım ve Yöntemler", Çukurova Üniversitesi Sosyal Bilimler Enstitüsü Dergisi, 18 (2): 458-480.

https://www.ntv.com.tr/galeri/yasam/en-buyuk-10-cevre-felaketi,qRjTxCun4Uuv2NgA_THz GQ/ Vu_FluNrmkqdcdtykqfNKQ, Access Date: 22.05.2018.

https://eleks.com.tr/tr/haber/detay/116/eca-serel-ile-surdurulebilir-bir-gelecek-ecologic-urun ler, Access Date: 09.06.2018.

https://www.capital.com.tr/capital-dergi/akilli-kimya/cevre-dostu-urun-donemi?sayfa=4, Access Date: 10.06.2018.

Hatice Aydın and Derya Fatma Biçer

Anti-Consumption for Environmental Sustainability

Introduction

Today's marketers attempt to behave in a profit and sales-oriented manner in the highly competitive environment. For this reason, they put more products on the market and encourage overconsumption. This tendency of the marketers can be reflected on today's consumers in different ways. The consumer is faced with a broad range of products and the danger of consumption. The continuous increase in product options resulting from quick technological developments can also increase the purchase of the consumers. In such an environment, overconsumption and anti-consumption are placed at either edge of consumer behaviors. In other words, some consumers experience non-satisfaction while some others tend to not consume (Kuzu and Özveren, 2011: 61). In the markets where these two situations are observed, the environmental degradation caused specially by the consumer community is tried to be solved especially through "anti-consumption" movement as well as sustainable consumption, thoughtful consumption, product sharing, voluntary simplicity movement and collective consumption (Kiraci, 2017: 195).

The tendency toward anti-consumption can be interpreted as a reaction. Participants of this movement refuse to consume because of many psychological, social and personal reasons rather than a necessity. It is possible to claim that one of the most significant reasons of this movement is environmental degradation. Human related environmental degradation that gives harm to the ecological balance decreases nature's potential to meet human needs. As a result, energy resources are quickly depleted. Depletion of energy resources is one of the issues that will jeopardize the sustainability of human beings and pave the way for "anti-consumption" movements. In general terms, anti-consumption can be defined as "acts or processes in which humans distribute their commodity to others for consumption and/or take something from them for consumption" (Belk, 2007: 126). Anti-consumption is a consumption bulk where products/services, natural and personal resources are voluntarily shared and an equal exchange relation is observed (Kiraci, 2017: 53).

It is considered that humans will become more motivated to adopt an anti-consumption attitude more when they are made more aware of the environmental problems (Chen and Chai, 2010). Environmental consciousness has become a significant issue in terms of shaping anti-consumption behaviors or tendencies of anti-consumers. In addition, it can be said that such actions as voluntary simple life (Cherrier and Murray, 2007), slow fashion (Fletcher 2007), negative emotions of consumers (Sandlin and Callahan, 2009), ethnocentrism and boycotts (Yener et al., 2016), collective consumption or sharing economy (Ozanne and Ballantine, 2010) may create anti-consumption behavior.

In recent years when movements to prevent environmental degradation have increased, it is possible to claim that the anti-consumers have stopped to be a member of a small or niche consumer group, but become the forerunners of a big trend in the market. Because of its significance, this section provides general information on motivations that support anti-consumption for environmental sustainability, classifies anti-consumption behaviors and explains some of them. In accordance with the provided information, suggestions are made.

1 Literature Review

With anti-consumption, personal, social and international aims are tried to be achieved. Preserving the continuity of life, being healthy and free, increasing life quality, not to be deceived by the system and being an essentialist are some of the personal aims. Socializing with anti-consumers and consumers can be expressed as a social aim. Enabling the economic and social justice and leaving a livable world to the next generations can be considered an international or social aim (Başcı, 2015). With anti-consumption, it is aimed to ensure environmental sustainability by means of "non-consumption" or "anti-consumption".

1.1 Environmental Sustainability

Sustainability means replacing non-renewable resources with renewable ones (Schaefer and Crane, 2005: 76). Sustainable consumption, on the other hand, was defined as "ensuring the consumption of general products and services without putting the needs of future generation under risk" (OECD, 2008: 11). Sustainable consumption aims to preserve today's life standards without consuming those of future generations. In this respect, the approach of sustainable consumption implies the usage of renewable natural resources of the environment and human actions. Sustainable consumption refers to a consumption process where the direction of investments is in compliance with technological equipment and corporate change and that meets the needs and expectations of potential humans of the future so that the natural resources can be preserved for future generations (Barde, 1990: 158). In order to ensure sustainable consumption, it is required to obey some environmental, economic and social principles. Decreasing all kinds of harmful substances and ensuring the recycle of wastes refer to environmental principles. Investing in ethical ways, supporting the equality of nations and local economies reflects economic principles. Ensuring social equality, cultural and social integration imply social principles (Sakınç, 2006: 4).

The solution for fulfilling economic, social and especially environmental principles of sustainable consumption can be to adopt an anti-consumption behavior. Many studies trace the reasons of product and brand prevention to sustainable life style (Black and Cherrie 2010: 439). Sustainable consumption can be associated with anti-consumption as it involves more activities in order to protect the environment. In short, it is possible to benefit from anti-consumption activities for sustainable consumption. In this respect, as a recent environmental trend, anti-consumption has gained a significant role in environmental sustainability (Cherrier et al., 2011: 1758). Thus, it is needed to know the notion of anti-consumption.

1.2 The Notion of Anti-Consumption

Anti-consumption can be defined as showing resistance and limiting wastage (Albinson et al., 2010: 414), simplifying life style (Ballantine and Creery, 2010), non-consumption (Shaw and Moraes, 2009) and voluntary simplicity (Shaw and Newholm, 2002). Anti-consumption activates people against consumption that does not bring happiness, but creates non-satisfaction. Accordingly, happiness does not mean possessing more or expensive things. A product is worth buying on the condition that it provides benefit or pleasure to the buyer (Kuzu and Özveren, 2011: 63). Otherwise, it will activate the desire of "non-consumption".

Anti-consumption involves not only general actions that are against consumption, but also those oriented toward more specific aims such as product/brand, company or nation (Chatzidakis and Lee, 2013). For instance, a consumer can adopt anti-consumption behavior toward environmentally-harmful products/brands that are not in compliance with his/her ideology (Sandıkcı and Ekici, 2009). Departing from this point of view, anti-consumption can be defined as "a resistance against a culture of consumption and the marketing of mass-produced meanings" (Penaloza and Price, 1993: 123), and considered a sub-branch of political and ideological consumption (Cherrier and Murray, 2007). In this respect, it can be expressed as boycotting (Kozinets and Handelman, 2004). In terms of not consuming a specific brand, it can be interpreted as brand avoidance (Lee et al., 2009a) and brand rejection (Sandikci and Ekici, 2009). As one of the general definitions, it can also be defined as voluntary simplicity (Shaw and Newholm, 2002). As far as these definitions are concerned, it is believed that the notion of anti-consumption is vague. In addition, there is not a clear-cut explanation in terms of its practice and attitude nature. In other words, it is not possible to define it exactly either as an "attitude" or "practice" (Cherrier, 2009).

In order to understand the notion of anti-consumption better, it is important not to ignore some issues. One of them is the necessity of non-consumption, decreasing consumption or selective consumption behaviors in order to talk about anti-consumption. The other issue is that it is required to stop or decrease consumption because of impersonal environmental problems, economic inequality, social discrimination or unethical marketing, or consume in a selective way because of similar reasons. Finally, it is important for the person adopting anti-consumption attitude to explain his/her reason for choosing to do so. The behavior of purchasing is required to be carried out in a rational way. When these issues are evaluated in a holistic manner, anti-consumption can briefly be defined as "the non-consumption, reduction-of-consumption or selective-consumption act that has a rational link to a societal and systemic problem on the local and/or global scene" (Basci, 2014: 162). This notion does not completely imply conscious consumption. It does not only refer to ethical, sustainable or public policies. Anti-consumption focuses on the requirements of a qualified consumption or the reasons of avoiding consumption. From the perspective of marketing, anti-consumption is not considered a natural economic threat, but an opportunity to gain information about ourselves, our products, practices and society (Lee et al., 2009b: 145). Categorizations made for clarifying this notion for which no exact definition has been provided can be guiding.

2 Classifying Anti-Consumption

Anti-consumption does not only focus on consumer's rejection of overconsumption, but it also emphasizes consumer actions for some specific aims such as product or brand, company or even country. In this respect, anti-consumption consumers change now from choosers group to active consumers (Chatzidakis and Lee, 2013). Departing from this change, many researchers evaluated anti-consumers in different categories (Kozinets and Handelman, 2004). One of the most common categories in the literature is considered to belong to Iyer and Muncy (2009).

According to Iyer and Mucy (2009), consumers can refuse consumption because of social and personal reasons. According to the consumers that reject consumption because of social reasons, consumption is not a simple purchase.

This consumer group believes that the society will develop to a great extent as long as consumption decreases. For them, the effect of consumption on society and its wealth is bigger. The consumers consuming for personal reasons try to not consume. For instance, they prefer not to consume in order to prevent pain or negative emotion that they will experience after purchasing or because they do not find it ethical (Iyer and Muncy, 2009). To sum up, the ones from general group keep themselves away from all kinds of consumption or all products and brands. The others from special groups, on the other hand, avoid consuming the symbolized brands or products. Departing from general and special groups, Iyer and Muncy dealt with four types of anti-consumer groups: They are global impact consumers, market activists, anti-loyal consumers and simplifiers (Iyer and Muncy, 2009).

Global impact consumers refer to the group that resists general consumption because of social reasons. They believe that there is strong social inequality and hence the current ecosystem cannot deal with the current overconsumption. In order to provide benefit to the society and environment, they want to decrease all consumption. Market activists comprise the group that resists obtaining special products and brands. These consumers blame specific brand or products for their illegal or immoral practices that give harm to the environment. They believe that they need to boycott these products and brands. Anti-loyal consumers represent the group whose resistance against consumption cannot be motivated because of social concerns. The consumers of this group cannot fulfil the behavior of purchasing because of not only their experience, but also the negative image of the mentioned product and brand. Simplifiers refer to the group resisting general consumption and thinking that they need to decrease all consumption. They believe that consumption keeps them away from more important life goals and that it is not a significant resource for happiness. For this reason, they resist consumption with the aim of having an easier, but more meaningful lives.

The behavior of anti-consumption cannot be independent from the society. For this reason, it would not be correct to classify anti-consumption by making a comparison between personal and social reasons. Anti-loyalty consumer classification is considered wrong because of similar reasons. It is thought that it would be beneficial to talk about a different classification, considering the criticism made for the classification provided by Iyer and Muncy (2009). Basci (2014: 166) also made a different classification. These groups include full-time reducing group that refuses overconsumption culture that has become a mass culture, that consumes less and/or uses the resources in the most efficient way; anti-consumerist boycotting group that keeps themselves away from specific products and brands in the long or short term with the aim of punishing the businesses; anti-consumerist boycotting group that purchases the products and brands preferred socially; interactive anti-consumption group that carries out anti-consumption activities in a collective manner and self-producing group that aims to produce goods and services in the market instead of benefiting from them.

As far as these classifications are concerned, it is easy to conclude that the notion of anti-consumption is quite complex. However, when we consider all of the classifications and definitions, it is possible to classify all of the activities aiming to minimize all practices of wastage that will give harm to environment and affect economic balance negatively. In accordance with the classifications and definitions provided above, it would contribute to the literature to talk about some principles or motivations.

3 Some Motivations to Support Anti-Consumption

In an environment where overconsumption is active, the ultimate aim of marketing is to enable sustainable consumption. It is believed that sustainability will be achieved by resisting against conspicuous, luxurious, over and wrong consumption as well as wasting. This resistance may create "anti-consumption" behaviors. Considering both the provided definitions and the classifications, there exist many motivations that can support anti-consumption behaviors. As far as environmental principles of anti-consumption are considered, it is possible to observe simple life. In terms of economic and social principles, on the other hand, boycotting consumption motivations are observed. For instance, in political consumption activities, consumers boycott the brands and products that are not in compliance with their own political ideologies (Sandıkcı ve Ekici, 2009) while they boycott the organizations that affect the society in a negative manner in anti-globalization analysis (Klein et al., 2004). However, not only voluntary simple life or boycott, but also many other motivations are considered significant for anti-consumption. In this respect, it is possible to talk about environmental consciousness, slow fashion, consumer guilt, collective consumption and the economy of sharing. As is seen, the motivations behind anti-consumption behavior are realized in various ways (Pentina and Amos, 2011).

3.1 Environmental Consciousness

In general terms, environmental consciousness can be defined as the concern level of the effects of the threats the environment faces upon future generations (Diamantopolous et al., 2003). The increase in environmental consciousness has resulted from the increase in the importance given to environmental problems.

Otherwise, the environmental problems may jeopardize the being and life quality of future generation and even cause some species to become extinct due to the disrupted balance of the nature. Such environmental problems as ozone layer spoil or global warning that causes a threat to the whole world increase the awareness of human beings. Those having environmental consciousness prefer more environment friendly products, consume less or never for the sustainability of natural resources while shaping their life styles and product choices that they will purchase, considering the negative effects of their behavior and choices upon the environment (Black and Cherrier, 2010). In this respect, environmental consciousness is considered one of the motivations behind anti-consumption that is believed to become a guide to enable sustainable consumption.

3.2 Voluntary Simplicity

As opposed to the consumption that does not bring satisfaction or peace, a group of consumers try to live in a simpler manner (Kuzu and Özveren, 2011: 70). This preference can be interpreted as "voluntary simplicity" movement. Voluntary simplicity can be defined as a social and individual reaction to consumer culture, materialist consumption and their harmful trends (Craig- Lees and Hill, 2002). Voluntary simplicity implying departure from consumer culture enables individuals to become aware of personal, environmental degradation and its negative outcomes, and help them to guarantee their well-being in the future by changing their behaviors. As this notion means less consumption and less material dependence, the life style of these people is highly significant in terms of the sustainability of natural resources (Cherrier, 2009). Voluntary simplifiers refuse to purchase the materials that do not improve their happiness level (Elgin, 1981) as well as the products that do not comply with their identity (Craig-Lees and Hill, 2002). In this respect, these followers of voluntary simplicity adopt an anti-consumption lifestyle with the aims of avoiding material addiction, using less resource and contributing to decreasing environmental and social effects of consumption culture (Black and Cherrier, 2010).

3.3 Slow Fashion

Many firms decrease the consumption of products or brands through re-use or recycling and hence attempt to minimize the environmental and social effects of consumption in fashion industry (Mangır, 2016). In this sense, they can support disposition and prevent wastage through slow fashion. People must understand that the continuous consumption of fashion will not bring much benefit to the society and it is unnecessary to buy new clothes every six weeks. In this way, anti-consumption can be supported with slow fashion and hence sustainability can be achieved. In order to talk about the sustainability of fashion, some features need to be fulfilled (Black and Cherrier, 2010). These features involve downscaling the scale of production, limiting products, decreasing the negative effect of production on the environment, paying labor wages fairly, establishing strong relations with suppliers and decreasing overconsumption by offering qualified products.

3.4 Ethnocentrism and Boycotting

Consumer ethnocentrism implies developing a prejudice against foreign products by prioritizing local products. The ethnocentric people can display a negative attitude because of potential harms that they can cause to the local economy in terms of foreign product or brands as well as job loss. The inequality and dependence caused by global firms may lead the ethnocentrics to adopt a negative attitude toward the products of such firms and prefer socially responsible firms and their products (Huneke, 2005). Another notion related to ethnocentrism is boycotts. They are anti-consumption actions happening as a result of dissatisfaction (Al Shebil, Rasheed and Al-Shammari, 2011).

The boycotting persons prefer environmentally sensitive alternatives by displaying infidel behaviors against the products and/or firms that give harm to the society and environment. Boycott suggests an action that is done when consumers do not approve the product or corporate behavior of a company, which can also be defined as an anti-consumption behavior (Taşçıoğlu and Yener, 2017: 56).

3.5 Consumer Guilt

Emotions can have an effect on consumer's rejection of consumption or their sensitivity. Such a self-conscious emotion as guilt can have a role in arranging the self (Antonetti and Maklan, 2014: 118). It has been seen that guilt affects individuals' abilities to control their decisions and enables them to reach their aims (Tangney and Tracy 2012). As far as the relation between guilt and environmental sustainability relation is considered, it can be said that the retailers can benefit from guilt in advertisements for controlling ethical consumption behaviors and enabling sustainable consumption (Steenhaut and Van Kenhove, 2005). As a result, environmental degradation can be prevented by being associated with such negative emotions as guilt (Mayer and Frantz, 2004).

3.6 Collaborative Consumption or Sharing Economy

As one of the outcomes of ownership-based economic system, overconsumption leads to environmental destructions. Sharing the products with many individuals in such an environment enables the waste of resources. Sharing means common use of the resources in joint tenancy with no thought of gain (Ozanne and Ballantine, 2010: 486). Sharing economy, on the other hand, is used synonymously with the notion of collective consumption that is expressed with the sentence "what is mine belongs to you" (Botsman and Rogers, 2010). Collective consumption suggests the spread of traditional sharing, exchange, borrowing, renting, gift-giving methods in public (Botsman and Rogers, 2010). In the literature on consumers, collective consumption is evaluated as various viewpoints and tendencies that involve sharing (Belk, 2014) borrowing (Jenkins et al., 2014), donating (Strahilevitz and Myers, 1998), second-hand market and sustainable consumption (Young, Hwang, McDonald and Oates, 2010) and even adopting an anti-consumption attitude (Ozanne and Ballantine, 2010). The notion of collective consumption is evaluated by Belk (2014) as one dimension of sharing behavior while Botsman and Rogers (2010 models it as an umbrella term that also includes the sharing behavior. According to Bardhi and Eckhardt (2012), however, it is seen as two main types of access-based consumption. There does not exist a certain answer for whether these two notions overlap or which one includes the other, and the notion of "sharing" is often confused with the notion of "collective consumption".

Conclusion

Under today's market conditions, supports for creating awareness about environment and the protection of world resources have been increasing. Because of the increased awareness on environmental sustainability and relevant issues, this study aims to put forth anti-consumption and the motivations in the market. According to the literature presented in this study, environment consciousness is considered one of the most significant motivations behind anti-consumption. It is possible to say that the ones with high environment awareness are more probable to support the general anti-consumption by considering the balance of the nature and the scarcity of the resources. The consumers adopting the voluntary simplicity as a life style ascribe their well-being to giving less harm to the environment. Some businesses or consumers in fashion industry believe that they can adopt an anti-consumption attitude and support sustainability by means of slow fashion movement. However, under today's conditions, having environment consciousness, supporting voluntary simplicity and following slow fashion movement is not considered enough to react against consumption. Especially in recent years, some consumers have adopted anti-consumption by developing negative attitudes against specific products and brands, boycotting with ethnocentric behaviors while some others share or consume collectively.

Considering the offered classifications, it is obvious that more classifications must be done related to the consumer group, and there still exists some gaps in this area. Especially the sharing economy and boycotts take attention as two different motivations behind anti-consumption. As far as the scarcity of world resources is considered, it is expected that the issues regarded by anti-consumers will contribute to providing ethically appropriate goods and services. The marketing experts can enable sustainability by learning the motivations behind consumers' environment friendly choices and non-consumption tendencies. They must keep in mind that negative emotions, ethnocentrism and boycott, collective consumption and sharing may also be considered among the motivations. Considering the motivations dealt in this study, the firms can increase the environment consciousness in an effective way by means of various media tools. As the avoidance of ethnocentric people to purchase the products of environmentally harmful firms will affect their market share and profit negatively, the firms need to be more sensitive in terms of ethnic values. It is significant that this tendency and its forerunners must have an increasing effect in the market, and the companies must adopt anti-consumption motivations in order to find out new opportunities.

References

Al Shebil, Saleh, Rasheed, Abdul A., and Al-Shammari, Hussam, (2011) "Coping with Boycotts: An Analysis and Framework", Journal of Management & Organization, 17: 383–397.

Albinsson, P. A., Wolf, M., and Kopf, D. A. (2010) "Anti-Consumption in East Germany: Consumer Resistance to Hyperconsumption", Journal of Consumer Behavior, 9(6): 412–425.

Antonetti, P., ve Maklan, S. (2014) "Feelings that Make a Difference: How Guilt and Pride Convince Consumers of the Effectiveness of Sustainable Consumption Choices", Journal of Business Ethics, 124(1): 117–134.

Ballantine P. W., and Creery S. (2010) "The Consumption and Disposition Behavior of Voluntary Simplifiers", Journal of Consumer Behavior, 9(1): 45–56.

Barde, J. P. (1990) "The Path to Sustainable Development", The OECD Observer, 164: 33.

Bardhi, F., ve Eckhardt, G. M. (2012) "Access-Based Consumption: The Case of Car Sharing", Journal of Consumer Research, 39: 881–898.

Basci, E. (2014) "A Revisited Concept of Anti-Consumption for Marketing", International Journal of Business and Social Science, 5(7(1)): 160–168.

Başcı, E. (2015) "Pazarlama ve Tüketim Toplumuna Eleştirel Bir Bakış: Tüketim Karşıtlığına İlişkin Nitel Bir Model", Yayınlanmamış Doktora Tezi, Anadolu Üniversitesi Sosyal Bilimler Enstitüsü, Eskişehir.

Belk, R. (2007) "Why not Share Rather Than Own?, The Annals of the American Academy of Political and Social Science", 611(1): 126–140.

Belk, R. (2014) "You are What You can Access: Sharing and Collaborative Consumption Online", Journal of Business Research, 67(8): 1595–1600.

Black, I. R., and Cherrier, H. (2010) "Anti-Consumption as part of Living a Sustainable Lifestyle: Daily Practices, Contextual Motivations and Subjective Values", Journal of Consumer Behavior, 9(12): 437–453.

Botsman, R., ve Rogers, R. (2010) "What's Mine is Yours: The Rise of Collaborative Consumption", Harper Business, New York.

Chatzidakis, A., ve Lee, M. S. W. (2013) "Anti-Consumption as the Study of Reasons Against", Journal of Macromarketing, 62(2): 145–147.

Chen, T. B., and Chai, L. T. (2010) "Attitude towards the Environment and Green Products: Consumers Perspective", Management Science and Engineering, 4(2): 27–39.

Cherrier, H. (2009) "Anti-Consumption Discourses and Consumer Resistant-Identities", Journal of Business Research, 62 (2): 181–90.

Cherrier, H., and Murray, J. B. (2007) "Reflexive Dispossession and the Self: Constructing a Processual Theory of Identity", Consumption Markets & Culture, 10(1): 1–29.

Cherrier, H., Black, I. R., and Lee, M. (2011) "Intentional Non-Consumption for Sustainability: Consumer Resistance and/or Anti-Consumption?", European Journal of Marketing, 45(11/12): 1757–1767.

Craig-Lees, M. and Hill, H. C. (2002) "Understanding Voluntary Simplifiers", Psychology & Marketing 19(2): 187–210.

Diamantopoulos, A., Schlegelmilch, B. B., Sinkovics, R. R., and Bohlen, G. M. (2003) "Can Socio-Demographics Still Play a Role in Profiling Green Consumers? A Review of the Evidence and an Empirical Investigation", Journal of Business Research, 56(6): 465–480.

Elgin D. (1981) "Voluntary Simplicity: Toward a Way of Life that is Outwardly Simple, Inwardly Rich", Morrow, New York.

Fletcher, K. (2007) "Slow Fashion", The Ecologist, 1.

Fournier S. (1998) "Consumer Resistance: Societal Motivations, Consumer Manifestations, and Implications in the Marketing Domain", Advances in Consumer Research 25: 88– 90.

Huneke, M. E. (2005) "The Face of the Un-Consumer: An Empirical Examination of the Practice of Voluntary Simplicity in the United States", Psychology and Marketing, 22 (7): 1–38.

Iyer, R., and Muncy, J. A. (2009) "Purpose and Object of Anti-Consumption", Journal of Business Research, 62(2): 160–168.

Jenkins, R., Molesworth, M., ve Scullion, R. (2014) "The Messy Social Lives of Objects: Inter-Personal Borrowing and the Ambiguity of Possession and Ownership", Journal of Consumer Behavior, 13(2): 131–139.

Kiraci, H. (2017) "Tatlıyı Yemek mi Yoksa Paylaşmak mı Tatlı? Paylaşım ve Ortak Tüketim Davranışı Üzerine Kuramsal Bir İnceleme", İşletme Araştırmaları Dergisi, 9(1): 194–220.

Klein J. G., Smith C. N., and Andrew J. (2004) "Why we Boycott: Consumer Motivations for Boycott Participation", Journal of Marketing, 68(3): 92–109.

Kozinets, R. V., and Handelman, J. M. (2004) "Adversaries of Consumption: Consumer Movements, Activism, and Ideology", Journal of Consumer Research, 31(3): 691–704.

Kuzu, A., ve Özveren, H. (2011) "Tüketilen Tüketici", SAÜ Fen Edebiyat Dergisi, 13(1): 61–72.

Lee M. S. W., Motion, J., and Conroy, D. (2009a) "Anti-Consumption and Brand Avoidance". Journal of Business Research, 62: 169–80.

Lee M. S. W., Fernandez K. V., and Hyman M.R. (2009b) "Anti-Consumption: An Overview and Research Agenda", Journal of Business Research 62(2): 145–147.

Lee, M. S., Seifert, M., and Cherrier, H. (2017) "Anti-Consumption and Governance in the Global Fashion Industry: Transparency is Key", In Governing Corporate Social Responsibility in the Apparel Industry after Rana Plaza, Palgrave Macmillan, New York, 147–174.

Mayer, F., and Frantz, C. (2004) "The Connectedness to Nature Scale: A Measure of Individuals' Feeling in Community with Nature", Journal of Environmental Psychology, 24(4): 503–515.

Mangir, A. F. (2016) "Sürdürülebilir Kalkınma için Yavaş ve Hızlı Moda", Selçuk Üniversitesi Sosyal Bilimler Meslek Yüksek Okulu Dergisi, 19: 143–154.

OECD. (2008), Promoting Sustainable Consumption: Good Practising in OECD Contries, OECD Publishing, USA.

Ozanne, L.K., ve Ballantine, P.W. (2010) "Sharing as a Form of Anti-Consumption? An Examination of Toy Library Users", Journal of Consumer Behavior, 9(6): 485–498.

Penaloza, L., ve Price, L.L. (1993) "Consumer Resistance: a Conceptual Review", Advances in Consumer Research, 20: 123–128.

Pentina, I., and Amos, C. (2011) "The Freegan Phenomenon: Anti-Consumption or Consumer Resistance?" European Journal of Marketing, 45(11/12): 1768–1778.

Sakınç, E. (2006) "Sürdürülebilirlik Bağlamında Mimaride Güneş Enerjisi Etken Sistemlerin Tasarım Öğesi Olarak Değerlendirlmesine Yönelik Bir Yaklaşım", Yıldız Teknik Üniversitesi, Fen Bilimleri Enstitüsü, Doktora Tezi, İstanbul.

Sandıkcı, O., and Ekici A. (2009) "Politically Motivated Brand Rejection", Journal of Business Research, 62(2): 208–217.

Sandlin, J. A., and Callahan, J. L. (2009) "Deviance, Dissonance, and Détournement: Culture Jammersuse of Emotion in Consumer Resistance", Journal of Consumer Culture, 9(1): 79–115.

Schaefer, A., and Crane, A. (2005) "Adressing Sustainability and Consumption", Journal of Macromarketing, 25(1): 76–92.

Shaw D., and Moraes C. (2009) "Voluntary Simplicity: An Exploration of Market Interactions", International, Journal of Consumer Studies, 33(2): 215–223.

Shaw, D., and Newholm, T. (2002) "Voluntary Simplicity and the Ethics of Consumption", Psychology & Marketing, 19(2): 167–85.

Steenhaut, S., and Van Kenhove, P. (2005) "Relationship Commitment and Ethical Consumer Behavior in a Retail Setting: The Case of Receiving too much Change at the Checkout", Journal of Business Ethics, 56(4): 335–353.

Strahilevitz, M., and Myers, J. G. (1998). "Donations to Charity as Purchase Incentives: How Well They Work May Depend on What You Are Trying to Sell", Journal of Consumer Research, 24(4), 434–446.

Tangney, J. P., and Tracy, J. L. (2012) "The Self-Conscious Emotions", In M. Leary and J. P. Tangney (Eds.), Handbook of Self and Identity (2nd ed., pp. 446–480), The Guilford Press, New York.

Taşçıoğlu, M., and Yener, D. (2017) "Tüketicilerin Boykotlara Karşı Tutumlarına Yönelik Bir Araştırma: Menşei Ülke ve Sürdürülebilirliğin Etkileri", Akademik Sosyal Araştırmalar Dergisi, 5(61): 54–67.

Yener, D., Dursun, T., and Oskaybaş, K. (2016) "Determinants that Affect Consumers'boycotts Participation", Akademik Sosyal Araştırmalar Dergisi, 4(33): 61–75.

Young, W., Hwang, K., McDonald, S., and Oates, C. J. (2010). "Sustainable Consumption: Green Consumer Behaviour When Purchasing Products", Sustainable Development, 18(1), 20–31.

Mutlu Uygun

The Green Brand Equity in terms of Environment and Sustainability

Introduction

The significant awareness on environment and sustainability that has started to develop in the society because of environmental pollution and waste of resources resulting from the increasing industrial activities in the world (Chen, 2008: 531; Chen, 2011: 384) has brought up the subject to the priority agenda of the enterprises again with its current concepts. In addition to legal regulations, the enterprises are forced to carry out sustainable and environmental friendly practices that regard the needs of future generations so that they can develop effective responses to changing consumer demands (Chen and Chai, 2010: 28). Turning it into an opportunity becomes strategically important, and green, ecological or sustainable marketing used in similar meanings in the literature implies a continuously increasing value for most of the sector. With the spread of environmental problems since the early 1990s, the increase in environmental attitudes of consumers foregrounds innovative environmental friendly marketing activities and business models that are in compliance with green tendencies (Chang and Chen, 2014: 1755). It is remarkable that a serious discussion ground has been established regarding the expected necessity for these practices to become integrated with all of the strategies of the enterprises in the near future. It is often emphasized that the enterprises are affected seriously by green consumer behaviors, and it is discussed that 70 % of the consumers pay attention to environmental issues in their shopping behaviors (Wagner, 2005: 1).

In traditional terms, green marketing includes all kinds of marketing processes and activities that involve development, differentiation, pricing, promotion and distribution of environmental friendly products and services (Chen and Chang, 2012: 503); while in its current meaning, it encompasses of all the enterprises activities designed to fulfill the needs and wants of consumers with minimum detrimental impact on the natural environment so that consumers' environmental friendly behaviors or attitudes can be stimulated and sustained (Park et al., 2010: 321). As promises regarding environmental and sustainability provide a green competition advantage (Chen and Chang, 2013: 529), even the enterprises that do not offer environmental friendly products nowadays attempt to develop a green corporate image (Bekk et al., 2016: 1728) and position themselves in this direction. It is possible to talk about such driving forces behind green marketing as "compliance with environmental pressures, obtaining competitive advantage, improving corporate images, seeking new markets or opportunities, and enhancing product value" (Grubor and Milovanov, 2017: 81) and hence improving profitability.

The products and services that satisfy environmental concerns of consumers make it possible to take the supports of environmentalist consumers (Chen, 2010: 307), create a preference to buy green products among other alternatives and even to be willing to pay more for them (Konuk et al., 2015: 586). For this reason, it is striking that the enterprises have started to prioritize consumer-oriented, integrated green marketing activities, but not all them have enough capacity for these practices (Chang and Chen, 2014: 1754), or receive the desired returns from the investments that they have made. Among the reasons behind it, the striking ones refer to the production of environmental friendly products that do not go beyond the functional performance of traditional ones, the insufficiencies in environmental friendly brand management as well as the inability to eliminate consumer skepticism (Ng et al., 2014: 204). An enterprise can turn itself into a completely green concept only with the appropriate strategies at organizational and product level (Butt et al., 2017: 508). Strategies at organizational level require efficient green management practices, while product-level ones involve developing products that have a minimum negative environmental impact. In the literature, it is understood that green marketing has been examined in the scope of two main viewpoints as "behaviorist and brand-oriented". Even though these two viewpoints are interconnected; in behaviorist approach, green consumption is examined by predicating on main variables of consumer behavior, while in brand-based approach, the way to ease the consumer's motivation for becoming a green consumer by means of the values of specific brands is explored. The brand perspective is relatively new in green marketing field, and the studies in this field have increased in recent years. In this respect, "brand equity" has started to attract noticeable attention in terms of research and practice.

In today's era when environmentalism and sustainability are realized as increasing values, strengthening green brand equity is both an obligation and an opportunity the enterprises need to discover. Grubor and Milovanov (2017: 81), argue that there exist strong relations between environmental friendly brand equity, competition advantage, brand strength and sustainable practices of an enterprise. Accordingly, it is possible to claim that environmental friendly (eco-friendly) brands with high values are an indispensable component of a sustainable marketing strategy. Therefore, understanding consumer-oriented green brand equity is vital in terms of a successful green brand management and obtaining the expected returns. However, although it is notable that brand equity has often been examined in the field of general marketing, it is also seen that this topic has been neglected in terms of green marketing, which has become a significant academic study area in the last thirty years, and it has not been sufficiently studied. As a result, this study has aimed to deal with green brand equity, the factors determining green brand equity, the potential effects of high green brand equity on enterprises and to discuss green brand management accordingly.

1 Green Brand Equity and Potential Effects

Brands that have been used by producers for centuries to differentiate their products from the others are critically important for the success with the difference that they create in the minds of consumers (Jung and Sung, 2008: 24; Pappu et al., 2005: 143). Nowadays, brands represent strong exchange tools, and are accepted as one of the main capitals (Kim and Kim, 2005: 549). Brands define the source of a product, determine the responsibility of a specific producer and most importantly have subjective meanings for consumers. As a result of this, brands ensure brevity and simplicity in product decisions. When consumers get acquainted with a brand and are knowledgeable about it, they do not need a long information processing. Brands provide consumers with unique and personal meanings that make their daily activities easier and enrich their lives (Keller, 2013: 34–35). Thus, the mutual relations between brands and consumers draw attention, and consumers determine the development and success of brands while brands affect and direct consumer behaviors. Brands become a way for consumers to express their interests, attitudes, preferences and general personalities or identities (Erdil, 2013: 14). Consumers can accept brands as an important component of daily life, prioritize them in their preferences and can defend the ideas hidden in the philosophy of their favorite brand in a strong manner. The enterprises having successful brands followed by loyal consumer groups may have the power to create a change or even a transformation in the lifestyle, value system, attitudes and behaviors of consumers (Grubor and Milovanov, 2017: 78). For this reason, creating an effective brand community in the society is accepted as a valuable asset.

"Brand equity", having been accepted as one of the most strategic sources of competitive advantage and sustainability in recent years (Kim and Kim, 2005: 549; Vogel et al., 2008: 98), has taken great interest in the fields of academic and practice. Brand equity is a benefit or value added to a product or service with the name of its brand (Yoo and Donthu, 2001: 1). This value is a relational structure that occurs when compared to rival brands (Ng et al., 2014: 206). Brand equity can be dealt within "financial and consumer" perspectives. Financial brand equity is the value based on the measurement of data on financial, accounting and store levels including the increased cash flows of a brand (Keller, 1993: 1). In this value, consumers' cognitive and behaviorist attitudes toward brands are neglected (Yoo and Donthu, 2001: 2). For the purposes of this study, customer-based brand equity is defined as a set of brand assets and liabilities associated with the name, terminology, logo or emblem of a brand whose value created with a product or service directed toward the brands or customers of enterprises can be increased or decreased (Aaker, 1991: 103). This is an intangible and implicit value in the inner nature of a well-known brand name and indicates consumer's preference, attitude and purchase behavior toward a specific brand (Grubor and Milovanov, 2017: 81). As brand equity is more based on subjective qualities perceived by consumers than objective characteristics of economic offering, two brands with the same objective features can imply different values for consumers. Therefore, brand equity refers to relative advantages and/or disadvantages of an economic offering compared to an objectively similar or even same brand (Bekk et al., 2016: 1728). Accordingly, brand equity is a set of the assets developed between the features of a brand and the benefits perceived by its customers (Chang and Chen, 2014: 1758; Keller, 1993: 2). Brand equity can contribute to increase in income and profit, decrease in cost (Keller, 1993; Kim and Kim, 2005), improve in business volume, to put high price and to increase in cash flows and consumer satisfaction (Pappu and Quester, 2006), desired purchase tendencies and behaviors, brand expansion and price flexibility (Wang et al., 2008).

Brand equity is also significant for environmental, sustainable and green entities of a brand. Green brand equity, derived from the notion of brand equity by Chen (2010: 310), is defined as "a set of brand assets and liabilities about green commitments and environmental concerns linked to a brand, its name and symbol that add to or subtract from the value provided by a product or service". Environmental factors that add value to the brand can also increase its green equity and make the economic offering detached from its objective environmental features. Though the objective green values obtained from rival two economic offerings are the same, one of them can carry a relatively higher green value because of subjective green image assets attributed to the brand for consumers (Bekk et al., 2016: 1728) and enable competitive advantage by positioning enterprises' products in a different way in the market (Butt et al., 2017: 511). Green brand equity develops strong emotional ties among consumers, supports positive attitudes toward these products and brands and hence strengthens public awareness as well as contributing to a strong financial structure for the future (Grubor and Milovanov, 2017: 81).

In order to benefit from green brand equity, it is necessary to get informed about its correlation with attitudes and behaviors. Green brand equity and its determiners are effective in brand attitudes of consumers (Bekk et al., 2016), brand loyalty (Kang and Hur, 2012; Konuk et al., 2015), positive suggestion behavior (Word of Mouth / WOM) (Bekk et al., 2016; Konuk, 2015) and willingness to pay higher price (Konuk et al., 2015; Ng et al., 2014). Green brand equity creates a positive mental image, which also affects brand attitude and loyalty. A positive attitude toward a brand occurs when subjective image of consumers is in compliance with brand image. Brand emotion, on the other hand, affects attitude and behavior-based loyalty of consumers (Chaudhuri and Holbrook, 2001: 84). Green brand equity can affect positive word-of-mouth communication indirectly by affecting brand attitudes of consumers.

Green brands need to be ecological (minimizing the negative effect on natural environment), fair (preventing marketing promotion directed toward unsustainable social practices) and economic (promoting long-term economic development) characteristics (Grubor and Milovanov, 2017: 80). The question of what enterprises should do for a strong green brand equity from which they will obtain a return that they desire requires dealing with the indicators of green brand equity and its probable results in a holistic manner.

2 Factors Determining Green Brand Equity

It is notable that various factors that affect green brand equity are examined in the previous studies. These factors include green brand image, green brand awareness, green brand trust, green perceived quality and green perceived risk, green brand affect, green brand perceived value etc. The correlation of all these factors with green brand equity is explained in the following section.

2.1 Effect of Green Brand Image on Green Brand Equity

Brand image is one of the most important factors that contribute to brand equity. Brand image playing important roles in the markets where it is difficult to differentiate products or services based on tangible quality features refer to consumers' mental picturing of the brand of an enterprise (Cretu and Brodie, 2007: 232). Brand image is consumer perceptions and preferences for a brand, measured by the various types of brand associations held in memory. Brand associations may be either brand attributes or benefits. Brand attributes are those descriptive features that characterize a product or service while brand benefits are the personal value and meaning that consumers attach to the product or service attributes (Keller, 2013: 77). Although brand associations come in many forms, Keller (2013: 549) states that it can usefully distinguish between product-related or performance-related versus non-product-related or imagery-related attributes. A useful distinction with benefits is between functional (intrinsic product advantages), symbolic (extrinsic product advantages) or experiential (product consumption advantages) benefits. Some of these attribute and benefit associations may be more rational or cognitive in nature; others more emotional or affective. Park et al. (1986) also argue that brand image involves functional, symbolic and experiential benefits.

Studies (Baran et al., 2017; Bekk et al., 2016; Butt et al., 2017; Chen, 2010; Ng et al., 2014; Pechyiam and Jaroenwanit, 2014) show that there is a positive correlation between green brand image and green brand equity. Green brand image can be defined as a whole range of impressions, conceptions and apprehensions toward a brand in the customers' memory which is correlated to the sustainability and eco-friendly concerns (Ng et al., 2014: 205) or as the a set of brand perception (Chen, 2010: 309). Therefore, the priority duty for green brand image is to create positive perceptions in consumers toward green brands (First and Khetriwal, 2010: 91; Mourad and Ahmed, 2012: 516). Creating an effective green brand image depends upon the ability to direct his/her positive environmental emotions, perceptions, attitudes and behaviors toward the brand of the consumer (Chen, 2008: 541). In this respect, rather than green products purchased by consumers, the brands that improve the products gain importance (Butt et al., 2017: 510). Positive green brand image has an effect on developing consumer satisfaction and creating a difference against the rivals and increases consumer trust by decreasing the perceived risks (Chang and Chen, 2014: 1757).

The relationship between general attitudes of consumers toward green products or brands and brand image (Butt et al., 2017: 516) indicates that consumers need to increase their interest in environmental values and hence strengthen their general attitude toward green products and brands and show the tendency of enterprisers to invest in creating positive perception regarding their own green brands. The enterprises can unite their resources with their environmentalist partners for national or international communication campaign that will support to strengthen consumer concern in environmental values and general attitudes toward green products and hence become integrated in consumer environmental education.

2.2 Effect of Green Brand Awareness on Green Brand Equity

Brand awareness is another factor that affects green brand equity. Brand equity is realized when consumer's brand awareness is high and she/he has strong, positive and unique brand assets in his/her mind (Keller, 2013: 73). Being an indicator of a reputation, brand awareness is the probability of a consumer's being conscious of the existence and assets of a brand. Brand awareness consists of brand recognition and recall dimensions. Brand recognition is the ability of a consumer to recognize a brand from its clues while brand recall refers to the ability of a consumer to differentiate a brand within a specific product group (Keller, 1993: 3; Yoo and Donthu, 2001: 3). Green brand awareness, on the other hand, refers to consumer's recognition and recall of the environmental friendly nature of a brand (Chang and Chen, 2014: 1756).

High brand awareness implies the power of the brand and increases market performance (Erdem and Swait, 1998: 138). Brand awareness can also improve perceived brand equity and affect consumer's perceived risk and trust because of the familiarity with the brand (Aaker, 1996: 114). As consumers generally tend to purchase well-known brands with which they are familiar, high brand awareness can increase the preference rate of a brand as well as customer loyalty (Keller, 1993: 3). For this reason, brand awareness is one of the key components of brand strategy and has a significant role in creating brand equity. These arguments can be also valid for green brands. The study conducted by Chang and Chen (2014) supports this idea.

2.3 Effect of Green Brand Trust on Green Brand Equity

In the literature, it is often emphasized that there exists a positive correlation between brand trust and brand equity (Delgado-Ballester and Munuera-Aleman, 2005; Erdem and Swait, 2004). Brand trust refers to the average consumer's awareness of the ability, reliability and responsibility of a brand (Ganesan, 1994: 3) and the willingness to rely on the ability of the brand to perform its stated functions (Chaudhuri and Holbrook, 2001: 82). Green brand trust, on the other hand, can be defined as the willingness of a consumer developed according to environmental performance of a brand (Chen, 2010: 311) as well as the tendency of consumers to believe that they keep their word (Chaudhuri and Holbrook, 2001: 85) in terms of green performance. Brand trust can also be interpreted as temporal experiential learning of a consumer (Kang and Hur, 2012: 308) and it involves the knowledge and experience resulted from direct or indirect contact of the consumer with the brand.

Consumer's high trust sense in a brand comes with a positive attitude and can decrease the perceived risks by reducing the costs of information processing and eliminating uncertainties in purchase decisions and consumption (Ng et al., 2014: 206). Brand trust also affects perceived brand value, which is a determinant for customer loyalty (Matzler et al., 2008: 156). The studies (Bekk et al., 2016; Butt et al., 2017; Chen, 2010; Kang and Hur, 2012; Konuk et al., 2015; Ng et al., 2014; Pechyiam and Jaroenwanit, 2014) show that green brand trust is one of the most important determinants of green brand equity. Brand trust is also significant for enterprises to reflect an environmental friendly image. Therefore, green brand trust can mediate the relationship between brand image and brand equity (Butt et al., 2017: 512; Chen, 2010: 311).

In the context of green marketing it is difficult to manage consumers' trust. Nowadays, it is often observed that there exists a gap between environmental activities of enterprises and consumer perceptions. The main reason behind this is the confusion caused by the brands that provide exaggerated or misleading information to the consumers regarding their environmental assets by means of practices called "greenwashing" (Parguel et al., 2011: 16). Because of the positive effects of a brand's environmental promises, enterprises can attempt to use it in their communication strategies even though this brand does not sometimes have environmental friendly qualities. The studies (Chen et al., 2013 etc.) indicate that greenwashing has direct negative effects and indirect negative effects also through green trust and green satisfaction on green consumer behaviors. It is possible to mention a similar relationship between greenwashing and brand equity. The proper relationship between the green brand and its consumers can only be achieved by grounding communication strategy on reality. This is especially much more important in terms of today's conditions under which consumers are driven by exaggerated and misleading green information. The success of a green brand depends upon the success of persuading consumers of green claims. It is important not to exaggerate in the promotional practices of green arguments, but just focus on informing consumers (Danciu, 2015: 55). Promises of the enterprises regarding green brands should be based on a reasonable, realist, consistent and correct communication. Foregrounding social welfare in communication can also increase the possibility of acceptance and the generation of expected reactions (Konuk et al., 2015: 594).

2.4 Effects of Green Quality and Risk Perceptions on Green Brand Equity

As perceived high quality of brands can decrease knowledge cost in product decisions (Erdem and Swait, 1998: 131) and perceived risks (Keller, 2013: 35), it is accepted as one of the most significant determinants of brand equity (Pappu and Quester, 2006: 10). Apart from some technical qualities, perceived quality refers to general quality or superiority in a number of specific features (Sweeney et al., 1999: 80) of a product, service or brand that are in compliance with the expectations of consumer compared to its alternatives (Keller, 2013: 187). Green perceived quality can be defined as a customer's general judgment about environmental superiority of a product, service or brand (Chang and Chen, 2014: 1755). Perceived quality can direct consumers toward a specific brand (Delafrooz and Goli, 2015: 4) by making it differentiate from its rival ones (Aaker, 1996: 109; Keller, 1993: 5), decrease the costs of managing customers, increase the purchase volume, enable higher pricing (Ng et al., 2014: 204) and create a positive word-of-mouth communication effect (Chang and Chen, 2014: 1758).

Perceived risk, on the other hand, is a combination of uncertainties and negative consequences (Stone and Gronhaug, 1993: 40) and a structure including physical, financial, psychological, performance and social risks. In perceived risk theory, it is assumed that consumers are more tended to first minimize perceived risk rather than maximize the expected positive outcome. Perceived high risk leads the consumer to avoid purchasing. The increasing environmental trends can increase perceived risks of consumers further. Green perceived risk can be defined as the expectation of negative environmental consequences associated with purchase behavior (Chen and Chang, 2012: 506). The studies (Bekk et al., 2016; Chang and Chen, 2014) show that green perceived quality and green perceived risk have direct effects on green brand equity. In addition, high green perceived quality positively affects green brand image, green brand trust and green perceived brand value and contributes indirectly to the development of green brand equity.

Some consumers can perceive green products with lower quality (Konuk et al., 2015: 594). Therefore, in the product development process, enterprises should not ignore the ability of products to solve the main problems of consumers and be persuasive about the fact that green products have the same and even higher quality standards.

2.5 Effect of Green Brand Affect on Green Brand Equity

The brand is a unique mixture of functional and emotional features perceived by consumers as an added value, a unique experience and a fulfilled promise (Lynch and Chernatony, 2004: 404). Brands can serve as symbolic devices, allowing consumers to project their self-image. Certain brands are associated with certain types of people and thus reflect different values or traits. Consuming such products is a means by which consumers can communicate to others – or even to themselves – the type of person they are or would like to be (Keller, 2013: 34). It is often emphasized that brands provide functional, symbolic and experiential benefits to consumers (Uygun and Akın, 2012). For this reason, the basic foundation behind brand equity is that the power of a brand is hidden in the mind and heart of consumers (Keller, 2013: 69). Accordingly, brand affect is one of the significant determinants of brand equity. Brand affect is defined as a brand's potential to elicit a positive emotional response in the average consumer as a result of its use (Chaudhuri and Holbrook, 2001: 82). Green brand affect, on the other hand, can be defined as the potential of a green brand usage to elicit a positive emotional response in a typical consumer. Environmental friendly brands that elicit positive emotions in a consumer encourage behavioral and attitudinal loyalty as assets with high green brand equity (Kang and Hur, 2012: 308). Consumers turn toward green brands that are in compliance with their subjective images, they are more likely to buy these brands again, and they join in environmental friendly products to a great extent. Kang and Hur (2012) determined that green brand affect that results from the interaction with environmental friendly qualities has a significant role in the development of green brand equity. Even though there exist lots of studies on general brand experiences in the literature, it is understood that the studies on green brand experience including green affects are not sufficient and more studies are crucially required in this area.

An enterprise needs to use both functional and emotional strategies for positioning green brand. The green brand positioning strategy, which is based on functional features, aims to create brand assets by providing information about environmental friendly products and their characteristics. In this respect, the product should be grounded with its environmental advantages compared to its traditional rival ones or refer to production processes, product use or product elimination. The functional features alone cannot be a motivating factor for the purchase of the brand. As they are based on rational purchasing decisions, they can even be imitated easily and decrease the flexibility of brand diversification. In emotional green brand positioning strategy, on the other hand, it assumed that these emotions are the key for establishing a bond with consumers. The emotional bond with consumers can create a loyalty that cannot be enabled with the features of products. This bond is based upon the principle of causing the environmentalist consumers to feel well by contributing to developing a common good environment, enabling them to express their identities through the socially visible consumption of green brands by displaying their environmental consciousness both for themselves and to the others as well as the emotion that they experience in their contact with the environment (Danciu, 2015: 55–57). For this reason, a persuasive communication based on consumers' motivation, including the mentioned components and reaching to the minds and hearts of consumers is required.

2.6 Effect of Green Brand Perceived Value on Green Brand Equity

Perceived value can generally be defined as the perception of difference between the benefit obtained from a product, service or brand and the compromises given during that process (Sweeney et al., 1999: 79), or as the evaluation regarding net benefit of a product, service or brand that a consumer expects to obtain or perceives its purchase (Chen and Chang, 2012: 505). Perceived brand value includes emotional components in addition to functional and rational ones (Wang et al., 2004: 171). Green perceived value, on the other hand, can be defined as the general evaluation between what is obtained and compromised regarding the net benefit of a product, service or brand based on a consumer's environmental desires, sustainable expectations and green needs (Chen and Chang, 2012: 506). The studies (Hartmann and Apaolaza-Ibanez, 2012; Malik, 2012), indicate that consumers believe that environmental friendly products provide additional benefits compared to traditional ones. Consumers' tendency to purchase can be increased by developing a positive green brand perceived value (Ng et al., 2014: 207). However, it should not be forgotten that not only environmental features, but also the abilities of providing superior quality that ensure success are important qualities of a brand.

The core of a successful green brand is that it should reflect a unique and high value for consumers, have the originality and meaning that will stimulate emotional responses and invade in all aspects of the activities of the enterprises. In order to create a successful green brand, the enterprises need to be aware of future directions of ecological efforts. Therefore, green value should include ecological innovations based on co-creation with consumers and be supported with an appropriate eco-communication (Danciu, 2015: 52–54).

Conclusion

It has been thought that this study will contribute to dealing with the factors that affect brand equity and their potential outcomes through a holistic approach, clarifying the processes related to green brand equity and creating a framework that will be used in practice. The significant factors that affect green brand equity include green brand image, green brand awareness, green brand trust, green perceived quality and green perceived risk, green brand affect and green brand perceived value. It is clear that positive developments ensured in each of these variables for strengthening green brand equity will contribute to the developments in green brand equity, consumers' purchase decisions, brand attitude and brand loyalty, positive word-of-mouth communication and willingness to pay higher price.

The fact that today's consumers are more aware of green values makes it difficult to persuade them only with re-packed products that are named "green". The common feature of successful brands is to understand what is important in people's lives, how and in which direction the shared culture changes, how to direct it rather than watch and how to create brand integrity, and apply them in brand strategies. This requires radical changes in marketing policy and organizational culture. Green brand strategy involves product design, production, packaging and positioning of products as well as practices in the fields of product's contact with the target market. For this reason, creating a successful green brand is a multidimensional process, which requires integrating long-term environmental marketing practices functioning as a determinant in strategic planning process with general strategy, activities and value chain of the enterprise. The enterprises need to activate ideas and current practices that will encourage green activities, be consistent, educate consumers and create effects that will bring about desired changes. In this respect, uniqueness, innovation, co-creation of a sustainable value and correct communication criteria should be taken into consideration, adopted and put in process as the main perspective in which both cognitive and emotional factors are effective. In addition, it should not be forgotten that practice differences will come out in terms of product and service brands.

The fact that studies on green brand equity are quite new and the number of enterprises that adopt it in their business models is insufficient creates an opportunity for future studies. This study examined the variables dealt in different contexts in a holistic manner. In future studies, it can be possible to obtain beneficial information by testing all determinants and outcome variables together. It is understood that most of the studies on green brand equity have been carried out within the scope of electricity and electronic products. In the future studies, conducting research that deals with various product categories and sectors will be beneficial to produce generalizable information. In the studies, the dominance of quantitative approach is obvious. Qualitative research that will benefit especially from projective techniques can contribute to foregrounding subjective meanings that include covert motivations related to green product and brands, and to deepening the subject. In this way, it can be possible to understand the reasons why and how consumers choose green brands. The longitudinal studies based on time series that enable the observation of dynamic changes in the attitudes and behaviors toward green brands can provide beneficial approaches to understand the long-term effects of environmental concerns and values.

References

Aaker, D. A. (1991) "Managing Brand Equity: Capitalizing on the Value of a Brand Name", New York: The Free Press.

Aaker, D. A. (1996) "Building Strong Brands", New York: The Free Press.

Baran, A., Söylemez C. and Yurdakul, M. (2017) "Algılanan Yeşil Kalite, Algılanan Yeşil Risk ve Yeşil Marka İmajının Yeşil Marka Değeri Üzerindeki Etkisinde Yeşil Güvenin Aracılık Rolü", Uluslararası Yönetim İktisat ve İşletme Dergisi, 13(Special Issue): 1–11. 1–11.

Bekk, M., Sporrle, M., Hedjasile, R. and Kers-Chreiter, R. (2016) "Greening the Competitive Advantage: Antecedents and Consequences of Green Brand Equity", Quality & Quantity, 50: 1727–1746.

Butt, M. M., Mushtaq, S., Afzal, A., Khong, K. W., Ong, F. S. and Ng, P. F. (2017) "Integrating Behavioural and Branding Perspectives to Maximize Green Brand Equity: A Holistic Approach", Business Strategy and the Environment, 26: 507–520.

Chang, C.-H. and Chen, Y.-S. (2014) "Managing Green Brand Equity: The Perspective of Perceived Risk Theory", Quality & Quantity, 48: 1753–1768.

Chaudhuri, A. and Holbrook, M. B. (2001) "The Chain of Effects from Brand Trust and Brand Affect to Brand Performance: The Role of Brand Loyalty", Journal of Marketing, 65(2): 81–93.

Chen, T. B. and Chai, L. T. (2010) "Attitude towards the Environment and Green Products: Consumers' Perspective", Management Science and Engineering, 4(2): 27–39.

Chen, Y.-S. (2008) "The Driver of Green Innovation and Green Image-Green Core Competence", Journal of Business Ethics, 81(3): 531–543.

Chen, Y.-S. (2010) "The Drivers of Green Brand Equity: Green Brand Image, Green Satisfaction, and Green Trust", Journal of Business Ethics, 93: 307–319.

Chen, Y.-S. (2011) "Green Organizational Identity: Sources and Consequence", Management Decision, 49(3): 384–404.

Chen, Y.-S. and Chang, C.-H. (2012) "Enhance Green Purchase Intentions: The Roles of Green Perceived Value, Green Perceived Risk, and Green Trust", Management Decision, 50(3): 502–520.

Chen, Y.-S. and Chang, C.-H. (2013) "Enhance Environmental Commitments and Green Intangible Assets Toward Green Competitive Advantages: An Analysis of Structural Equation Modeling (SEM)", Quality & Quantity, 47(1): 529–543.

Cretu, A. E. and Brodie, R. J. (2007) "The Influence of Brand Image and Company Reputation where Manufacturers Market to Small Firms: A

Customer Value Perspective", Industrial Marketing Management, 36(2): 230–240.

Danciu, V. (2015) "Successful Green Branding, a New Shift in Brand Strategy: Why and How it Works", The Romanian Economic Journal, 56(XVIII): 47–64.

Delafrooz, N. and Goli, A. (2015) "The Factors Affecting the Green Brand Equity of Electronic Products: Green Marketing", Cogent Business & Management, 2: 1–12.

Delgado-Ballester, E. and Munuera-Aleman, J. L. (2005) "Does Brand Trust Matter to Brand Equity?", Journal of Product and Brand Management, 14(3): 187–196.

Erdem, T. and Swait, J. (1998) "Brand Equity as a Signaling Phenomenon", Journal of Consumer Psychology, 7(2): 131–157.

Erdem, T. and Swait, J. (2004) "Brand Credibility, Brand Consideration, and Choice", Journal of Consumer Research, 31: 191–198.

Erdil, T.S. (2013) "Strategic Brand Management Based on Sustainable-Oriented View: An Evaluation in Turkish Home Appliance Industry", Procedia-Social and Behavioral Sciences, 99: 122–132.

First, I. and Khetriwal, D. S. (2010) "Exploring the Relationship between Environmental Orientation and Brand Value: Is there Fire or Only Smoke?", Business Strategy and the Environment, 19(2): 90–103.

Ganesan, S. (1994) "Determinants of Long-Term Orientation in Buyer-Seller Relationships", Journal of Marketing, 58(2): 1–19.

Grubor, A. and Milovanov, O. (2017) "Brand Strategies in the Era of Sustainability", Interdisciplinary Description of Complex Systems, 15(1): 78–88.

Hartmann, P. and Apaolaza-Ibanez, V. (2012) "Consumer Attitude and Purchase Intention toward Green Energy Brands: The Roles of Psychological Benefits and Environmental Concern", Journal of Business Research, 65(9): 1254–1263.

Jung, J. and Sung, E. (2008) "Consumer-Based Brand Equity: Comparisons among Americans and South Koreans in the USA and South Koreans in Korea", Journal of Fashion Marketing and Management: An International Journal, 12(1): 24–35.

Kang, S. and Hur, W.-M. (2012) "Investigating the Antecedents of Green Brand Equity: A Sustainable Development Perspective", Corporate Social Responsibility and Environmental Management, 19: 306–316.

Keller, K. L. (1993) "Conceptualizing, Measuring, and Managing Customer-Based Brand Equity", Journal of Marketing, 57, 1–22.

Keller, K. L. (2013) "Strategic Brand Management: Building, Measuring, and Managing Brand Equity", Fourth Edition. England: Pearson Education Limited.

Kim, H.-B. and Kim, W.G. (2005) "The Relationship between Brand Equity and Firms Performance in Luxury Hotels and Chain Restaurants", Tourism Management, 26: 549–560.

Konuk, F. A., Rahman, S. U. and Salo, J. (2015) "Antecedents of Green Behavioral Intentions: A Cross-Country Study of Turkey, Finland and Pachistan", International Journal of Consumer Studies, 39: 586–596.

Lynch, J. and Chernatony, L. (2004) "The Power of Emotion: Brand Communication in Business-To-Business Markets", Journal of Brand Management, 11(5): 403–419.

Malik, S. U. (2012) "Customer Satisfaction, Perceived Service Quality and Mediating Role of Perceived Value", International Journal of Marketing Studies, 4(1): 68–76.

Matzler, K, Grabner-Krauter, S. and Bidmon, S. (2008) "Risk Aversion and Brand Loyalty: The Mediating Role of Brand Trust and Brand Affect", The Journal of Product and Brand Management, 17(3): 154–162.

Mourad, M. and Ahmed, Y.S. (2012) "Perception of Green Brand in an Emerging Innovative Market", European Journal of Innovation Management, 15(4): 514–537.

Ng, P. F., Butt, M. M., Khong, K. W. and Ong, F. S. (2014) "Antecedents of Green Brand Equity: An Integrated Approach", Journal of Business Ethics, 121: 203–215.

Pappu, R. and Quester, P. (2006) "Does Customer Satisfaction Lead to Improved Brand Equity? An Empirical Examination of Two Categories of Retail Brands", Journal of Product & Brand Management, 15(1): 4–14.

Pappu, R., Quester, P. G. and Cooksey, R. W. (2005) "Consumer-Based Brand Equity: Improving the Measurement-Empirical Evidence", Journal of Product & Brand Management, 14(3): 143–154.

Parguel, B., Benoit-Moreau, F. and Larceneux, F. (2011) "How Sustainability Ratings Might Deter 'Greenwashing: A Closer Look at Ethical Corporate Communication", Journal of Business Ethics, 102: 15–28.

Park, C. W., Jaworski, B. J. and MacInnis, D. J. (1986) "Strategic Brand Concept-Image Management", Journal of Marketing, 50(4): 135–145.

Park, J., Ko, E. and Kim, S. (2010) "Consumer Behavior in Green Marketing for Luxury Brand: A Cross-Cultural Study of US, Japan and Korea", Journal of Global Academy of Marketing, 20(4): 319–333.

Pechyiam, C. and Jaroenwanit, P. (2014) "The Factors Affecting Green Brand Equity of Electronic Products in Thailand", The Macrotheme, 9(3): 1–12.

Stone, R.N. and Gronhaug, K. (1993) "Perceived Risk: Further Considerations for the Marketing Discipline", European Journal of Marketing, 27(3): 39–50.

Sweeney, J.C., Soutar, G.N. and Johnson, L.W. (1999) "The Role of Perceived Risk in the Quality-Value Relationship: A Study in a Retail Environment", Journal of Retailing, 75(1): 77–105.

Uygun, M. and Akın, E. (2012) "Markaların İşlevsel, Sembolik ve Deneyimsel Yararlarına İlişkin Tüketici Değerlendirmelerinin İncelenmesi", Anadolu Üniversitesi Sosyal Bilimler Dergisi, 12(2): 107–122.

Vogel, V., Evanschitzky, H. and Ramaseshan, B. (2008) "Customer Equity Drivers and Future Sales", Journal of Marketing, 72: 98–108.

Wagner, S. A. (2005) "Understanding Green Consumer Behavioral: A Qualitative Cognitive Approach", New York: Taylor & Francis Group.

Wang, H., Wei, Y. and Yu, C. (2008) "Global Brand Equity Model: Combining Customer-Based with Product-Market Outcome Approaches", Journal of Product & Brand Management, 17: 305–316.

Wang, Y., Po Lo, H., Chi, R. and Yang, Y. (2004) "An Integrated Framework for Customer Value and Customer-Relationship Management Performance: A Customer-based Perspective from China", Managing Service Quality: An International Journal, 14(2/3): 169–182.

Yoo, B. and Donthu, N. (2001) "Developing and Validating a Multidimensional Consumer-Based Brand Equity Scale", Journal of Business Research, 52: 1–14.

Halim Tatlı and Beşir Koç

The Relationship between Environmental and Socio-Demographic Factors that Affect Consumers' Demand for Goods

Introduction

Non-environmentally friendly production and consumption behaviors cause environmental pollution, finally leading to global warming which is expressed as a major problem by the scientific community. Global warming and environmental pollution increase air and water pollution and lead to the formation of acid rain, reduced plant diversity, ozone layer depletion and climate change, finally leading to the depletion of earth's resources. Environmental problems threaten the whole world and are framed as global problems. Within the scope of the environmental policies introduced by many countries, legal amendments have been made both to promote eco-friendly production and to ensure that consumers gain eco-friendly consumption habits. Nowadays, promoting eco-friendly consumer behaviors is among the methods employed to ensure sustainable consumption. According to the rational choice theory proposed by neoclassical economists, human is homo economicus. In other words, we want to maximize utility when we demand any goods by taking account of the price of the goods and the income we have. However, the human beings are under the influence of their habits, emotions, social norms, moral behaviors, cognitive limitations and the environment when demanding goods (Jackson, 2005). Studies conducted in the field of behavioral economics in the last fifty years have shown that the extent to which individuals behave rationally in decision-making is limited (Simon 1957). In the context of behavioral economics, there are several studies revealing that the environmental factor is important and affects consumers' demand for goods (Shogren and Taylor, 2008; Gowdy, 2008, Yakita and Yamauchi, 2011).

Concerns that emerged following the emergence of the phenomenon of environmental degradation have led some firms to design environmentally friendly products and improve their environmental performance of packaging. Increasing the number of such production behaviors of firms depends on the behaviors of consumers. Because if environmental factors become important in changing consumer behavior, environmental awareness may also gain importance in the production behavior of firms. Natural resources which are limited on earth may be on the verge of running out as a result of environmental pollution. The way to ensure sustainable development and sustainable consumption is to protect the environment. In this sense, attention should be paid to the relationship between the consumers' behavior of buying goods and services and their environmental behavior. The link between consumers' environmental attitudes and behaviors was established by Diekmann and Preisendörfer (1992). When establishing this link, they developed the low-cost hypothesis. This hypothesis assumes that consumers perform a "cost-benefit" analysis from their own perspective while exhibiting environmentally-friendly behavior (Tatlı, 2017). A behavior is considered as a "low-cost" behavior if it does not require the individual to sacrifice his or her comfort or to pay money and if it is easy to do; while the opposite behaviors are considered as "high-cost" behaviors. Consumers' demands for goods and services result in a cost that comes along with a benefit. A part of this cost, which can be expressed as private cost, is the cost that the consumer pays and the other is the environmental costs that may arise as a result of the act of consumption which we can express as social cost. Any increase in social costs may lead to the loss of scarce resources on the earth. For this reason, analyzing consumers' interests and attitudes toward the environment while demanding goods and identifying such consumer behaviors are of high importance.

Our literature review shows that there are studies examining the relation between consumers' environmental attitude and decision about buying goods and services, consumers' environmental concern and the general relationship between environmental factors and socio-demographic factors. Hines et al. (1987), Zimmer et al. (1994), Chan (1996), Kilbourne and Beckmann (1998) and Barr et al. (2003) found that environmental factors are important in consumers' demands for goods and services. Similarly, some studies reported that environmental concerns directly affect consumers' decision to buy green products (Kim and Choi, 2005; Mostafa, 2009). Some studies (Gleim et al. 2013; Ritter et al. 2015) examined the decision to buy green products using the theory of planned behavior proposed by Ajzen (1985, 1991). Laroche et al. (2001) and Beckford et al. (2010) examined the relationship between environmental attitudes and consumers' decision to buy green products. For example, Laroche et al. (2001) reported that environmental attitudes were more important in decisions to buy green goods and services than environmental concern and environmental consciousness, and environmental attitudes of consumers encouraged them to pay higher price for green products. Similarly, Beckford et al. (2010) found that environmental attitude is an important factor in consumers' demand for green products. The result of the empirical analysis performed by Chekima et al. (2016) revealed that environmental attitude, eco-label and cultural value have a significant impact on green purchase intention.

It is seen that previous studies yielded different results regarding the relationship between environmental factors and socio-demographic factors. It is observed that socio-demographic factors may have both positive and negative significant relationship with buying green products. Environmental factors were found to be significantly affected by age (Shen and Saijo, 2008; Yau, 2012; Sang and Bekhet, 2015), by gender (Diamantopoulos et al. 2003; Ay et al. 2005; Yuan and Zuo, 2013; Zhao et al. 2014), by educational background (Babekoğlu, 2000; Diamantopoulos et al. 2003; Zarnikau, 2003; Armağan and Karatürk, 2014), by the number of children (Grunert, 1991), by the income level (Roe et al. 2001; Zarnikau, 2003; Bekhet and Al-alak, 2011) and by occupation (Yuan and Zuo, 2013; and Zhao et al. 2014). Some studies reported that there was no significant relationship between gender and environmental attitudes of consumers (Yau, 2012; Chekima et al. 2016). Similarly, there are also studies that found no significant relationship between marital status and environmental factors and variables (Neuman, 1986).

A general overview of the literature shows that the number of studies conducted at the micro level is higher. These studies are mostly questionnaire-based. Studies are mostly aimed at identifying the factors that influence the buying decisions of consumers. The studies on the relationship between environmental factors and socio-demographic factors are limited in number. Therefore, this study first classifies consumer behaviors based on the environmental factors that affect consumers' demand for goods and then examines the relationship between these classified factors and socio-demographic factors. Due to the contemporary nature of the issue studied, the diverse range of variables chosen and the availability of primary data obtained through questionnaires, we believe that this study will make a contribution to the field and attract the attention of policy makers once again to this issue.

1 Material and Method

This study aims is to identify the environmental factors that affect consumers' demand for good in the context of behavioral economics and to empirically determine the relationship of these factors with socio-demographic factors. To this end, a questionnaire was administered to the participants through face-to-face interview. The questionnaire consists of two sections. The first section included socioeconomic and demographic characteristics, whereas the second section included consumers' statements on whether the environment is considered to have an impact on their demand for goods. The study conducted by Üstündağlı and Güzeloğlu (2015) was taken as basis while preparing the questionnaire. A pilot questionnaire was administered to 15 participants. Following the pilot administration, a question was excluded from the questionnaire and it was put into its final form to be administered to the target population of the study. The questionnaires were administered to individuals through face-to-face interviews within a three-month period covering January-February-March 2016. They were distributed proportionally across the neighborhoods in the central district of Mardin. Besides, the questionnaires were administered to individuals chosen randomly in each neighborhood.

According to the data dated 31 December 2016 obtained from the Address-Based Population Registration System of the Turkish Statistical Institute, Mardin city center has a population of 163 thousand 725 people, which represents 21 % of the total population in the province. The target population of the study consists of the people from the Artuklu district. The minimum sample size to represent the target population was determined to be about 383 people at 1 % level of significance with a 5 % margin of error (https://www.surveysystem.com/sscalc.htm). Any questionnaire filled out incorrectly was excluded from the analysis. The analysis was performed on 382 questionnaires.

We used the Factor Analysis, the Kruskal Wallis Variance Analysis, the Mann-Whitney U test and Spearman's Rank-Order Correlation to analyze the data. We also used multiple comparison tests. The scale developed to identify the environmental factors that affect consumers' demand for goods was first subject to the factor analysis. Then the relationship of consumers with socio-demographic factors was tested using the four subscale scores produced by the factor analysis. Since the sampling size used in the study was greater than 30, we used the Single Sample Kolmogorov Smirnov Test and the Homogeneity of Variance Test to test whether each variable was distributed normally or not. As a result of these tests, we found that the data was not distributed normally. Therefore, we used non-parametric tests in the analysis. Averages ($\bar{X}$) were presented with standard deviation (S) and statistical significance was determined to be $p < 0.05$.

Fig. 1 shows the relationship among the factors that generally affect the demand for goods. Fig. 1 summarizes the design of the study. Accordingly, environmental preferences and attitudes and socioeconomic and demographic factors may have an impact on demands for goods separately. Moreover, socioeconomic and demographic factors may affect the demand for goods by affecting environmental preferences and attitudes (Fig. 1).

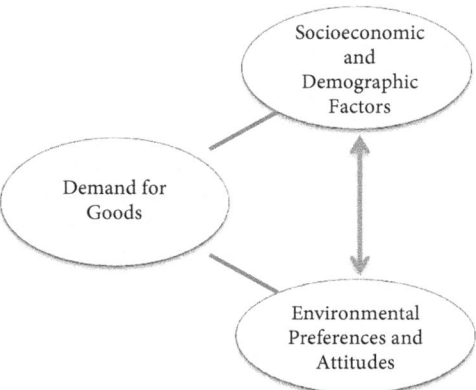

Fig. 1: The Relationship between the Factors that Affect the Demand for Goods in General. Source: Created by Authors

2 Research Findings

2.1 Sampling Characteristics

Before giving information about the analysis part of the study, we need to present the descriptive statistics in order to provide information about the sampling characteristics. Tab. 1 presents the socio-demographic characteristics that will identify the participants in the study.

Socio-demographic characteristics of the participants show that 58.9 % of them were female and 55.8 % of them were single/widow (Tab. 1). In terms of educational background, 37.4 % of the participants were high school graduates and 32.5 % of them had a bachelor's degree (Tab. 1). Similarly, Tab. 1 shows that the participants were mostly (62 %) aged 30 years or less. Finally, 38.5 % of the participants had a monthly income of 1401–2400 TL.

51.8 % of the participants stated that they did shopping 1–2 times a week and 39.8 % stated that they did shopping 3–4 times a week. These results show that the participants in the study sample are experienced in shopping, so they can answer the questions in the questionnaires correctly.

2.2 Factor Analysis

In order to interpret each statement in the subscales produced by the factor analysis, means and standard deviations of the participants' responses to the

Tab. 1: Socio-Demographic Characteristics of the Participants. Source: Created by Authors

Gender	Frequency	%	Marital Status	Frequency	%
Female	225	58.9	Single/Widow	213	55.8
Male	157	41.1	Married	169	44.2
Education	Frequency	%	Age	Frequency	%
Elementary School	21	5.5	30 and below	237	62.0
Middle School	50	13.1	31–40	69	18.1
High School	143	37.4	41–50	43	11.3
Associate Degree	39	10.2	51 and above	33	8.6
Bachelor's Degree	124	32.5			
Postgraduate	5	1.3			
Income	Frequency	%			
1400 and below	128	33.5			
1401–2400	147	38.5			
2401–3400	68	17.8			
3401 and above	39	10.2			
Total	382	100	Total	382	100

questions in the scale measuring the environmental perception in the demand for goods were given in Tab. 2.

Tab. 3 presents the distribution of the statements regarding the environmental attitude exhibited by the consumers while demanding goods. Given the findings obtained, it is seen that consumers definitely agreed with the following statements: "Unless everyone understands that the environment must be protected, next generations will suffer the consequences.", "Each individual should be educated about environmental protection" and "Global measures should be taken to mitigate environmental problems". This is because the mean score of these statements range between 4.21 and 5.00 (Tab. 2). In addition, the consumers agreed with the statement, "I would like to buy products in recyclable packages rather than those with non-recyclable packages". That is to say, consumers prefer products in recyclable packages in order to protect the environment while demanding goods. Similarly, consumers are concerned that the state must take a very active role in environmental protection (Tab. 2).

These statements are important in showing the desired results. However, the relationships between these statements and the level of these relationships should also be determined. Therefore, the statements and the responses of the participants were also factor analyzed.

Before performing the factor analysis, the appropriateness and adequacy of the data and sampling for the principal components analysis should be examined.

Tab. 2: The Statements Appear in the Scale and the Statistics of the Participants' Responses. Source: Created by Authors

Statements	$\bar{X}$	S
Each individual should be educated about environmental protection	4.366	0.801
The benefits of the prevention of environmental damages do not guarantee the expenses which are made for that purpose.	2.953	1.031
The importance of the environment is generally exaggerated.	3.796	1.065
I cannot personally help the deceleration of environmental devastation.	3.728	1.039
If it is possible, I buy the product which I consider that product is environment-friendly.	3.296	0.927
When I buy a product, I think about how that product affects the environment and other consumers.	2.901	0.922
I would like to buy products in recyclable packages rather than those with non-recyclable packages.	3.945	1.108
If the Central Government makes contribution to the environmental protection, the total effect would further increase.	3.762	0.974
If the Local Authorities makes contribution to the environmental protection, the total effect would further increase.	3.736	0.988
I use papers which are produced by recycling.	2.694	0.892
When I buy a product, I look into its ingredients to see whether it is environment-friendly.	2.741	0.957
Global measures should be taken to mitigate environmental problems.	4.335	0.895
Unless everyone understands that the environment must be protected, next generations will suffer the consequences	4.455	0.843
In order to consume fewer packages, I would like to buy bigger sized packages rather than the smaller sized packages of the same product.	2.997	1.083

For this purpose, we performed the Kaiser-Meyer-Olkin (KMO) and Bartlett's test of sphericity on the statements in the scale. Tab. 3 shows the results.

If the value of KMO (measure of sampling adequacy) is greater than 60 % and the Bartlett's test of sphericity is significant (0.05), then we can say that the data is suitable for the factor analysis. The results shown in Tab. 3 revealed that the values were sufficient to proceed with the factor analysis. Based on the results of these tests, the data were factor analyzed and four factors were identified. As a result of the factor analysis, 5 statements (5, 8, 12, 15 and 16) were excluded. Tab. 4 shows the results obtained from the exploratory factor analysis.

Consumers named these factors and the variables that constitute these factors as follows: These factors were identified to be A Consumer Who Has Internalized

Tab. 3: KMO and Bartlett Test of Sphericity. Source: Created by Authors

Kaiser-Mayer-Olkin (KMO) Sampling Measurement Value Sufficiency	0.654	
Bartlett's Test of Sphericity	Chi-Square	1311.398
	df	91
	Sig. 000	0.000

Environmental Consciousness, A Consumer Who Attaches Importance to the State's Role in Environmental Protection and A Passive Consumer with Low Environmental Concern. Tab. 4 presents the statements that constitute these factors, their factor loads, eigenvalues, variances and Cronbach's Alpha values. A comparison of the obtained Cronbach's Alpha value with the given ranges showed us that the scale had a sufficient level of reliability to be used for community surveys. This revealed that the scale was suitable for the analysis and the results were reliable.

- *A Consumer Who Has Internalized Environmental Consciousness:* This factor, which is ranked in the first place, consists of 5 statements and explains 17.852 % of the total variance. All of the statements making up this factor are related to awareness of environmental protection when demanding goods. Since these statements indicated that the participants who demanded goods were aware of the damages and positive impacts of their actions, this factor was named as A Consumer Who Has Internalized Environmental Consciousness. The eigenvalues and variances of the statements showed that the factor "A Consumer Who Has Internalized Environmental Consciousness" was the most important factor for the participants.

- *A Consumer with Environmental Concern:* This factor consists of 4 statements and explains 14.012 % of the total variance. It ranks second in terms of the extent to which it is found important by the consumers demanding goods. The statements making up this factor revealed consumers' concern for environmental protection and the measures for protecting the environment.

- *A Consumer Who Attaches Importance to the State's Role in Environmental Protection:* The statements in this factor, which explains 13.664 % of the total variance and ranks third in order of importance, show that the participants think the state must have an active role in environmental protection. Therefore, the factor was named as "A Consumer Who Attaches Importance to the State's Role in Environmental Protection". With the factor load of 0.882, the statement, "If local administrations contribute to environmental protection, the total impact will be greater." ranks first. This shows that consumers consider that local administrations are more important in protecting the environment.

Tab. 4: Sub-Dimensions Obtained through Factor Analysis. Source: Created by Authors

Statements Constituting Factors and Sub-Factors	Factor Loads	Eigen value	Variant %	Cronbach's Alpha
Factor 1: Consumer Who Has Internalized Environmental Consciousness		2.499	17.852	0.719
I use papers which are produced by recycling (13)	0.751			
When I buy a product, I look into its ingredients to see whether it is environment-friendly. (14)	0.749			
When I buy a product, I think about how that product affects the environment and other consumers. (7)	0.707			
If it is possible, I buy the product which I consider is environment-friendly. (6)	0.702			
In order to consume fewer packages, I would like to buy bigger sized packages rather than the smaller sized packages of the same product. (19)	0.478			
Factor 2: Consumer with Environmental Concern		1.962	14.012	0.622
Global measures should be taken to mitigate environmental problems. (17)	0.773			
Unless everyone understands that the environment must be protected, next generations will suffer the consequences (18)	0.738			
I would like to buy products in recyclable packages rather than those with non-recyclable packages. (9)	0.635			
Each individual should be educated about environmental protection. (1)	0.551			
Factor 3: A Consumer Who Attaches Importance to the State's Role in Environmental Protection		1.913	13.664	0.875
If the Local Authorities makes contribution to the environmental protection, the total effect would further increase. (11)	0.882			
If the Central Government makes contribution to the environmental protection, the total effect would further increase. (10)	0.869			

(continued on next page)

Tab. 4: (continued)

Statements Constituting Factors and Sub-Factors	Factor Loads	Eigen value	Variant %	Cronbach's Alpha
Factor 4: A Passive Consumer with Low Environmental Concern.		1.694	12.097	0.540
The importance of the environment is generally exaggerated. (3)	0.773			
I cannot personally help the deceleration of environmental devastation.(4)	0.692			
The benefits of the prevention of environmental damages do not guarantee the expenses which are made for that purpose. (2)	0.604			
Total			57.625	

Tab. 5: The Correlation between the Environmental Factors that Affect Consumers' Demand for Goods. Source: Created by Authors

	1	2	3	4
A Consumer Who Has Internalized Environmental Consciousness	1			
A Consumer with Environmental Concern	0.054	1		
A Consumer Who Attaches Importance to the State's Role in Environmental Protection	0.229*	0.127**	1	
A Passive Consumer with Low Environmental Concern.	-0.047	0.127**	0.072	1

*$p<0.001$, **$p<0.05$

- **A Passive Consumer with Low Environmental Concern:** The statements in this factor show negative and passive behaviors of consumers regarding environmental protection. Therefore, this factor was named as "A Passive Consumer with Low Environmental Concern". The factor explains 12.097 % of the total variance and ranks fourth. Given the factor loadings of these statements, it can be said that they have values close to each other.

Tab. 5 shows the correlation between the environmental factors that affect consumers' demand for goods. It was found that there was a weak and positive relationship between the average score of the scale "A Consumer Who Has Internalized Environmental Consciousness" and the average score of the scale "A Consumer Who Attaches Importance to the State's Role in Environmental Protection". Similarly, a weak and positive relationship was also found between the average score of the scale "A Consumer with Environmental Concern" and the average score of the scale "A Consumer Who Attaches Importance to the

State's Role in Environmental Protection". These results show that consumers want the state to be more effective in protecting the environment as their level of internalizing environmental consciousness increases.

2.3 Hypothesis Testing for the Relationship between Environmental and Demographic Factors that Affect the Demand for Goods

The Kruskal-Wallis H test was used to find out whether the ages of the consumers who participated in the study made any difference in the scale scores of the environmental factor and to test the statistical significance of this difference. Tab. 6 shows the results of the analysis.

Tab. 6: The Environmental Factors that Affect Consumers' Demand for Goods According to Age Groups. Source: Created by Authors

	Age	n	Mean rank	df	χ^2	P**	Significance difference
A Consumer Who Has Internalized Environmental Consciousness	30 and below	237	192.56	3	8.064	**0.045***	51 and above 41–50
	31–40	69	198.70				
	41–50	43	211.33				
	51 and above	33	143.02				
A Consumer with Environmental Concern	30 and below	237	189.69	3	1.423	0.700	No Difference
	31–40	69	201.02				
	41–50	43	198.37				
	51 and above	33	175.65				
A Consumer Who Attaches Importance to the State's Role in Environmental Protection	30 and below	237	204.06	3	10.804	**0.013***	31–40 and 30 and below
	31–40	69	163.72				
	41–50	43	193.86				
	51 and above	33	156.32				
A Passive Consumer with Low Environmental Concern.	30 and below	237	185.16	3	4.140	0.247	No Difference
	31–40	69	188.41				
	41–50	43	210.19				
	51 and above	33	219.17				

* $p<0.05$, ** Kruskal-Wallis Test

A significant difference (p<0.05) was found between the consumers of different ages regarding the behaviors of *A Consumer Who Has Internalized Environmental Consciousness* and *A Consumer Who Attaches Importance to the State's Role in Environmental Protection*. However, the difference was not found to be significant in terms of other consumer behaviors (p>0.05). When we examine the mean scores got by consumers of different ages regarding the behavior of *A Consumer Who Has Internalized Environmental Consciousness*, we can say that the fact that the score of consumers aged 51 years and above is lower than the score of those aged 41 to 50 years is a factor that has an impact on the difference between the groups. In other words, consumers aged 41 to 50 years might be said to have internalized environmental consciousness to a significantly more extent.

The analysis showed a statistically significant relationship (at a significance level of 1 %) between marital status of consumers and their scores on the behavior scale "*a consumer who attaches importance to the state's role in environmental protection*" (Tab. 7). In other words, there is a significant difference between the distributions of scores in terms of consumers' marital status. The mean ranks of consumers' marital status show that single/widow consumers have higher mean ranks. This finding shows that single/widow consumers attach more importance to the state's role in environmental protection.

Tab. 7: The Environmental Factors that Affect Consumers' Demand for Goods According to Marital Status. Source: Created by Authors

Factors	Marital status	n	Mean rank	Rank sum	U (???)	P^{**}
A Consumer Who Has Internalized Environmental Consciousness	Single/widow	213	193.90	41300.50	17487.50	0.634
	Married	169	188.48	31852.50		
A Consumer with Environmental Concern	Single/widow	213	192.41	40982.50	17805.50	0.857
	Married	169	190.36	32170.50		
A Consumer Who Attaches Importance to the State's Role in Environmental Protection	Single/widow	213	205.02	43669.50	15118.50	**0.007***
	Married	169	174.46	29483.50		
A Passive Consumer with Low Environmental Concern.	Single/widow	213	184.61	39321.50	16530.50	0.171
	Married	169	200.19	33831.50		

* p<0.01, **Mann Whitney U test

Based on the results of the analysis, there is no significant difference between consumers' gender and number of children they have and the distributions of scores on the scales regarding environmental factors that affect the demand for goods (p>0.05).

Tab. 8 shows the results of the analysis of the scores of consumers with different educational backgrounds on the scale regarding environmental factors that affect the demand for goods. The results of the analysis show that there is a statistically significant difference (at a significance level of 5 %) between the mean scores attained by consumers with different educational backgrounds on the behavior scale "a consumer who internalizes environmental consciousness". Based on the total scores on the scale in terms of consumers' education backgrounds, we can say that consumers with an associate or master's degree got the significantly highest mean scores. This shows that consumers with higher levels of education internalize environmental consciousness to a greater extent while demanding goods. The results of the multiple comparison analysis performed to find out the levels of education between which there was a difference in terms of consumers' mean scores show that there is a significant difference between the scores of consumers with a high school and postgraduate degree and the scores of those with a high school and bachelor's degree.

Besides, there is a significant difference (at a significance level of 5 %) between consumers' scores on the behavior scale "*A Consumer with Environmental Concern*" based on different educational backgrounds (Tab. 8). The mean scores of consumers with different educational backgrounds show that high school graduates got a significantly higher score on the scale. The results of the multiple comparison analysis performed to find out the levels of education between which there was a difference in terms of consumers' mean scores show that there is a significant difference (p=0.004) between the scores of consumers with a high school degree and those with a bachelor's degree. This finding of the study suggests that those with a high school degree are more concerned about the environment than those with a bachelor's degree. This result also shows that relatively younger consumers are more concerned about the environment.

Tab. 9 shows the results of the analysis of the scores of consumers with different income levels on the scale regarding environmental factors that affect the demand for goods. According to the results, there is a significant difference only between the behavior of *a consumer with environmental concern* and the income groups (p <0.05); however, there is no significant difference in terms of other scales (p> 0.05). In addition, the correlation analysis performed between the scale regarding the environmental factors that affect consumers' demand for good showed a positive relationship between income level and the behavior

Tab. 8: The Environmental Factors that Affect Consumers' Demand for Goods According to Consumers with Different Educational Backgrounds. Source: Created by Authors

Factors	Education	n	Mean Rank	df	χ^2	P***	Significance Difference
A Consumer Who Has Internalized Environmental Consciousness	Elementary School	21	186.71	5	8.064	**0.005***	High school-Bachelor's degree and High school postgraduate
	Middle School	50	173.59				
	High School	143	172.67				
	Associate Degree	39	205.44				
	Bachelor's Degree	124	211.88				
	Postgraduate	5	315.10				
A Consumer with Environmental Concern	Elementary School	21	153.76	5	1.423	**0.019****	Bachelor's degree – High School
	Middle School	50	186.47				
	High School	143	213.03				
	Associate Degree	39	208.49				
	Bachelor's Degree	124	170.51				
	Postgraduate	5	172.50				
A Consumer Who Attaches Importance to the State's Role in Environmental Protection	Elementary School	21	179.48	5	10.804	**0.043****	No differences between groups
	Middle School	50	167.69				
	High School	143	181.17				
	Associate Degree	39	180.44				
	Bachelor's Degree	124	218.16				
	Postgraduate	5	200.50				
A Passive Consumer with Low Environmental Concern.	Elementary School	21	217.29	5	4.140	0.158	No difference
	Middle School	50	204.63				
	High School	143	188.15				
	Associate Degree	39	216.92				
	Bachelor's Degree	124	175.75				
	Postgraduate	5	240.10				

* $p<0.01$, ** $p<0.05$, *** Kruskal-Wallis Test

Tab. 9: The Environmental Factors that Affect Consumers' Demand for Goods According to Different Income Levels. Source: Created by Authors

Factors	Age	n	Mean Rank	sd	χ^2	P**
A Consumer Who Has Internalized Environmental Consciousness	1400 and below	128	189.14	3	2.090	0.554
	1401–2400	147	185.26			
	2401–3400	68	197.63			
	3400 and above	39	212.06			
A Consumer with Environmental Concern	1400 and below	128	176.80	3	8.439	**0.038**
	1401–2400	147	208.48			
	2401–3400	68	173.22			
	3400 and above	39	207.60			
A Consumer Who Attaches Importance to the State's Role in Environmental Protection	1400 and below	128	207.04	3	6.049	0.109
	1401–2400	147	176.23			
	2401–3400	68	200.41			
	3400 and above	39	182.50			
A Passive Consumer with Low Environmental Concern.	1400 and below	128	186.53	3	2.492	0.477
	1401–2400	147	197.77			
	2401–3400	68	178.31			
	3400 and above	39	207.19			

* $p<0.05$, ** Kruskal-Wallis Test

of A Consumer with Environmental Concern (Correlation coefficient: 0.103, p=0.044, n=382). We can say that environmental concern increases as the income level increases.

Conclusion

This study aimed to determine the environmental factors that affect consumers' demand for goods and to reveal the relationship between these factors and the socio-demographic characteristic of consumers. The data obtained through a survey prepared for this purpose were analyzed using exploratory factor analysis. As a result of the factor analysis, four scales were obtained. These scales were named as follows: *A Consumer Who Has Internalized Environmental Consciousness, A Consumer Who Attaches Importance to the State's Role in Environmental Protection and A Passive Consumer with Low Environmental Concern.* Then, the relationships between the mean scores consumers got on these scales and the socio-demographic characteristics of the consumers were investigated using the Mann Whitney U Test and the Kruskal-Wallis Test.

At the end of the study, it was determined that consumers mostly exhibited the behavior of "*A Consumer Who Has Internalized Environmental Consciousness*" while demanding goods. They exhibited the behaviors of "A Consumer with Environmental Concern" and "A Consumer Who Attaches Importance to the State's Role in Environmental Protection". The behavior of "A Passive Consumer with Low Environmental Concern" was found to be the behavior exhibited by consumers to the least extent.

As a result of the tests carried out to determine the relationship between the factors that affect consumers' demand for goods and the socio-demographic characteristics of consumers, we obtained findings some of which were similar to the findings in the literature, while other were different. We found that there was a significant difference between the mean scores that consumers of different ages got on the scales titled "*A Consumer Who Has Internalized Environmental Consciousness*" and "*A Consumer Who Attaches Importance to the State's Role in Environmental Protection*". Accordingly, we can say that younger consumers have internalized environmental consciousness to a greater extent and attach more importance to the state's role in environmental protection. The results of some studies in the literature support these findings (Armağan and Karatürk, 2014; Zhao, 2014).

We found that there was a significant difference between the mean scores that consumers with different marital status got on the scale titled "*A Consumer Who Attaches Importance to the State's Role in Environmental Protection*". According to these findings, we can say that single consumers attach more importance to the state's role in environmental protection than married consumers.

There is no significant difference between consumers' gender and the number of children they have and the environmental factors that affect their demand for goods. This indicates that the environmental factors that affect consumers' demand for goods are not significant in terms of consumers' gender and the number of children they have. Similar to this finding, Yau (2012) and Chekima et al. (2016) did not found any significant relationship between marital status and the environmental factors and variables that affect purchase intention. Another empirical study reported that consumers' environmental attitudes did not change depending on the number of children they had (Diamantopoulos et al. 2003).

In this study, we found a significant relationship between the educational backgrounds of consumers and the behaviors of *A Consumer Who Has Internalized Environmental Consciousness, A Consumer with Environmental Concern and A Consumer Who Attaches Importance to the State's Role in Environmental Protection*. We also found that both consumers with a postgraduate degree and consumers with a bachelor's degree internalized environmental consciousness to a significantly greater extent than those with a high school degree. This finding shows that consumers who receive more education internalize environmental consciousness to a greater extent. As consumers' level of education increases, their environmental consciousness and their attitude toward demanding green products are positively affected (Straughan and Roberts, 1999). Zhao (2014) found that education level is related to environmental consciousness, environmental attitude and environmental concern. Similar to this study, the studies conducted by Üstündagli and Guzeloglu (2015) and Armagan and Karaturk (2014) found that consumers' education level was effective in the selection of environmentally friendly products. Such findings of other empirical studies support the findings of this study.

In this study, we observed that there was a statistically significant difference between the mean scores got by consumers with different income levels on the scale titled "a consumer with environmental concern." This means that high-income consumers are more concerned about the environment than low-income consumers. Similar to this finding, some studies reported that consumers' environmental consciousness and awareness increase and their buying behavior is affected as their income level increases (Zimmer et al. 1994; Bekhet and Al-alak, 2011; Yau, 2012). Moreover, consumers who demand goods more easily tolerate increases in the prices of green products as their income increases (Ay and Ecevit, 2005).

A general review of the results of the study indicates that there is a significant relationship between consumers' age, marital status, educational background and income level and the environmental factors that affect their demand for goods.

In conclusion, we recommended that policies should be developed to promote environmental protection and to ensure that consumers internalize environmental consciousness while demanding goods. The results of the study indicate that education is one of these policies. In order to create environmental awareness, individuals might be educated on buying goods with environmental consciousness at every stage of education. Courses on environmental consciousness might be offered at the primary, secondary and high school levels. Within the scope of undergraduate programs, courses on environmental economics might be offered. Such courses might be helpful for consumers to assume responsibility for the environmental damages that occur during the process of consumption and to internalize such environmental damages. Besides, we recommend that local administrations should provide young people with trainings and activities that promote the environment. We also recommend that companies should design products by taking environmental behaviors of consumers into account in order to increase their sales revenue.

References

Ajzen, I. (1985) "From Intentions to Actions: A Theory of Planned Behavior", In: Kuhl, J., Beckmann, J. (Eds.), Action Control: From Cognition to Behavior. Springer Verlag, New York.

Ajzen, I. (1991) "The Theory of Planned Behavior", Organizational Behavior and Human Decision Processes, 50(2): 179-211.

Armağan, E. ve Karatürk, H.E. (2014). "Yeşil Pazarlama Faaliyetleri Çerçevesinde Aydın Bölgesindeki Tüketicilerin Çevreye Duyarlı Ürünleri Kullanma Eğilimlerini Belirlemeye Yönelik Bir Araştırma", Organizasyon ve Yönetim Bilimleri Dergisi, 6(1): 1-17.

Ay, C. ve Ecevit, Z. (2005) "Çevre Bilinçli Tüketiciler", Akdeniz İİBF Dergisi, 10: 238-263.

Babekoğlu, Y. (2000) "Tüketicilerin Demografik Özellikleri ve Bireysel Tutumlarının Sorumlu Tüketim Davranışları Üzerindeki Etkisi", Doktora Tezi, Ankara Üniversitesi Fen Bilimleri Enstitüsü, Ankara.

Barr, S., Ford, N.J. ve Gilg, A. (2003) "Attitudes towards Recycling Household Waste in Exeter, Devon: Quantitative and Qualitative Approach", Local Environment, 8(4): 407-421.

Beckford, C.L., Jacobs, C., Williams, N. ve Nahdee, R. (2010) "Aboriginal Environmental Wisdom, Stewardship and Sustainability: Lessons from The Walpole Island First Nations, Ontario, Canada", The Journal of Environmental Education, 41(4): 239-248.

Bekhet, H.A. ve Al-alak, B.A. (2011) "Measuring E-Statement Quality Impact on Customer Satisfaction and Loyalty", International Journal of Electronic Finance, 5(4): 299-315.

Chan, T.S. (1996) "Concerns for Environmental Issues and Consumer Purchase Preferences: A Two Country Study", Journal of International Consumer Marketing, 9(1): 43-55.

Chekima B., Wafa, S.S., Wafa, S.K., Igau, O.A., Chekima, S. and Sondoh. S.L. (2016). "Examining Green Consumerism Motivational Drivers: Does Premium Price and Demographics Matter to Green Purchasing?", Journal of Cleaner Production, 112(4): 3436-3450.

Diamantopoulos, A., Schlegelmilch, B.B., Sinkovics, R.R. ve Bohlen, G.M. (2003) "Can Socio Demographics Still Play a role in Profiling Green Consumers? A Review of the Evidence and an Empirical Investigation", Journal of Business Research, 56(6): 465-480.

Diekmann, A. ve Preisendörfer, P. (1992). "Persönliches Umweltverhalten: Diskrepanzen zwischen Anspruch und Wirklichkeit", Kölner Zeitschrift für Soziologie und Sozialpsychologie, 40(2): 226-251.

Gleim, M.R., Smith, J., Andrews, D. and Cronin, J.J. (2013) "Against the Green: A Multimethod Examination of the Barriers to Green Consumption", Journal of Retailing, 89(1): 44–61

Gowdy, J.M. (2008) "Behavioral Economics and Climate Change Policy", Journal of Economic Behavior & Organization, 68(3–4): 632–644.

Grunert, S.C. (1991) "Everybody Seems Concerned About the Environment: But is this Concern Reflected in (Danish) Consumers' Food Choice?", Aarhus School of Business Working Paper Series, Aarhus.

Hines, J., Hungerford, H. ve Tomera, A. (1987) "Analysis and Syntheses of Research on Environmental Behaviour: A Meta-Analysis", Journal of Environmental Education, 18(2): 1–8.

Jackson, T. (2005) "Motivating Sustainable Consumption, a Review of Evidence on Consumer Behaviour and Behavioural Change", A Report to the Sustainable Development Research Network, University of Surrey, Guildford.

Kilbourne, W.E. ve Beckmann, S.C. (1998) "Review and Critical Assessment of Research on Marketing and The Environment", Journal of Marketing Management, 14(6): 513–532.

Kim, Y. and Choi, S.R. (2005) "Antecedents of Green Purchase Behaviour: An Examination of Collectivism, Environmental Concern and PCE", Advances in Consumer Research, 32(1): 592–599.

Laroche, M., Bergeron, J. ve Barbaro-Forleo, G. (2001) "Targeting Consumer Who are Willing to Pay More for Environmentally Friendly Products", Journal of Consumer Marketing, 18(6): 503–520.

Mostafa, M. (2009) "Shades of Green: A Psychographic Segmentation of the Green Consumer in Kuwait Using Self-Organizing Maps", Expert Systems with Applications, 36(8): 11030–11038.

Neuman, K. (1986) "Personal Values and Commitment to Energy Conservation", Environment and Behavior, 18(1): 53–74.

Ritter, A.M., Borchardt, M., Vaccaro, G.L.R., Pereira, G.M. ve Almeida, F. (2015) "Motivations for Promoting the Consumption of Green Products in an Emerging Country: Exploring Attitudes of Brazilian Consumers", Journal of Cleaner Production, 106(1): 507–520.

Roe, B., Teisl, M.F., Levy, A. ve Russell, M. (2001) "US Consumers' Willingness to Pay for Green Electricity", Energy Policy, 29(11): 917–25.

Sang, Y.N. ve Bekhet, H.A. (2015) "Modelling Electric Vehicle Usage Intentions: An Empirical Study in Malaysia", Journal of Cleaner Production, 92(1): 75–83.

Shen, J. ve Saijo, T. (2008) "Reexamining the Relations between Socio-Demographic Characteristics and Individual Environmental

Concern: Evidence from Shanghai Data", Journal of Environmental Psychology, 28(1): 42–50.

Shogren, J. ve Taylor, L. (2008) "On Behavioral-Environmental Economics", Review of Environmental Economics and Policy, 29(1): 26–44.

Simon, H.A. (1957). "Models of Man", New York: Wiley

Straughan, R. ve Roberts, J.A. (1999) "Environmental Segmentation Alternatives: A Look at Green Consumer Behavior in the New Millennium", Journal of Consumer Marketing, 16(6): 558–575.

Tatlı, H. (2017) "Tüketici Talep Davranışları Analizi, Hanehalkının Enerji Talebini Etkileyen Faktörler", Gazi kitapevi, Ankara.

Üstündağlı, E. ve Güzeloğlu, E. (2015) "Gençlerin Yeşil Tüketim Profili: Farkındalık, Tutum ve Davranış Pratiklerine Yönelik Analiz", Global Media Journal TR Edition, 5(10): 341–362

Yakita, A. ve Yamauchi, H. (2011) "Environmental Awareness and Environmental R&D Spillovers in Differentiated Duopoly", Research in Economics, 65: 137–143.

Yau, Y. (2012) "Eco-Labels and Willingness-To-Pay: A Hong Kong Study", Smart and Sustainable Built Environment, 1(3): 277–290.

Yuan, X. ve Zuo, J. (2013) "A Critical Assessment of the Higher Education for Sustainable Development from Students' Perspectives-a Chinese Study", Journal of Cleaner Production, 48: 108–115.

Zarnikau, J. (2003), "Consumer Demand for 'Green Power' and Energy Efficiency", Energy Policy, 31(15): 1661–1672.

Zhao, H., Wu, Y., Wang, Y. ve Zhu, X. (2014) "What Affects Green Consumer Behavior in China? A Case Study from Qingdao", Journal of Cleaner Production, 63: 143–161.

Zimmer, M.R., Stafford, T.F. ve Stafford, M.R. (1994) "Green Issues: Dimensions of Environmental Concern", Journal of Business Research, 30(1): 63–74.

Mehtap Çakmak Barsbay and Aytuğ Altın

Environment and Healthcare Sector: Current Debates on Sustainability

Introduction

Global changes such as rapid population growth, industrialization, urbanization and technological developments bring about some interventions on the environment made by humans. Environmental problems occur as a result of these human-induced interventions that disturb the natural balance. Environmental problems, which were grouped under air, land and water pollution, during the early days of industrialization, have increased and become more diversified as a result of the factors that emerged over time. Today, there are many global environmental problems, ranging from radioactive pollution, noise pollution and the loss of biodiversity to ozone layer depletion.

In Turkey, healthcare services include services of public nature provided either by the state or by the private sector under the guidance and supervision of the state. Hazardous materials used as required by the nature of healthcare services, wastes generated during the provision of healthcare services and the use of energy in healthcare services constitute the environmental dimension of healthcare services. The environmental awareness of environmental problems has brought the need to address the environmental dimension of healthcare services. In this sense, issues such as construction of energy efficient green buildings, hazardous materials management, waste management and construction of green hospitals have been focused on in the literature. This study serves the purpose of assessing the development of a sustainable healthcare sector within the framework of environmental impacts of this sector.

1 The Concept of Environment

The environment is the sum of all external conditions affecting the life, development and survival of an organism (UN, 1997). It is possible to define the environment as a special area where all living and non-living beings exist and with which a living unit or a community maintains a mutual relationship (Akdur, 2005: 14). The environment also refers to the place units that living beings are bounded with vital bonds and that they affect and are affected by. All of the external conditions that affect or are affected by living organisms constitute the environment (Görmez, 2010: 4).

In the Environment Law No. 2872, which forms the basis of the environmental legislation in Turkey, environment is defined as the biological, physical, social, economic and cultural medium in which the living beings continue their relation through all their lifetimes and interact mutually. In other words, environment is a system, in a given time, consisting of physical, chemical, biological and social factors that might have a direct or indirect, immediate or long-term effect on human activities and living organisms, and there is no space and time not covered by the environment (Keleş, Hamamcı and Çoban, 2012: 51).

The environment is divided into two types: natural and artificial environment. Natural environment refers to the environment unmodified by human intervention (Görmez, 2010: 4). Humans do not make any contribution to the formation of natural environment and find it readily available. Natural environment consists of two components: biotic and abiotic components. Whereas biotic components of the natural environment comprise of humans, plants and animals, abiotic components comprise of air, water, land, earth's layers and underground sources (Keleş et al., 2012: 54). Artificial environment refers to the environment created as a result of the interventions of humans since their appearance on earth (Görmez, 2010: 4). In other words, artificial environment is the environment that humans have created by their own hands using surface and underground sources based on their knowledge and cultural background (Keleş et al., 2012: 54). All values and assets that humans created using the natural environment are considered within the scope of artificial environment (Ertürk, 2012: 76).

2 Environmental Problems

In general, environmental problems refer to the negative impacts of the artificial environment on the natural environment (Ertürk, 2012: 4). Environmental problems also include hitches in the artificial environment and problems in both environments (Görmez, 2010: 5). Increasing population, developing industries and pollution which threaten the natural assets of countries have made environmental problems one of the most important problems of humanity in the early 21st century (Karacan, 2007: 340).

Air, land and water pollution constitute the main environmental problems facing humanity. In the strict sense, these are environmental problems and are called environmental pollution. In the broad sense, environmental problems refer to the problems that go beyond classical pollution in terms of size and effect (Görmez, 2010: 28–29).

Nature has been seen as an unlimited resource and it has been destroyed and polluted, which in the end has led to the emergence of environmental problems. The problems have increased to the point where they have no limits, forcing humans to take responsibility for eliminating these problems (Sipahi, 2010: 332). The current situation in terms of environmental problems shows that the stage where the problems can be solved by the efforts of single country has already been passed. Global environmental problems have become the most important problems for humanity (Yalçın, 2009: 289).

Environmental problems can be effective not only on a local scale, but also on a national, regional and global scale. In particular, global environmental problems affect all nations, and actions based on international cooperation are taken and legal arrangements are made to solve these problems.

Although many international agreements have been concluded since the Stockholm Conference where environmental issues were brought up for the first time at the international level, environmental problems such as environmental degradation, poverty, hunger, drought, deforestation, loss of biodiversity, climate change and global warming continue to grow throughout the world. Moreover, it is seen that the gap between developed and developing countries in this regard is going deeper (Sipahi, 2010: 332). Whereas environmental problems have an impact on countries at different levels, they have become important global problems that threaten the world in general.

2.1 Causes of Environmental Problems

The human-nature relationship has continuously deteriorated since the period where humans adopted a sedentary life and created an artificial environment. This deterioration is increasing due to rapid population growth, urbanization and industrialization, which cause human activities to concentrate in a certain area (Ertürk, 2012: 2).

Countries that have been developing rapidly since the 20th century have achieved social welfare and reached their economic development goals. However, infinite human needs have also made demands infinite, and new demands have led to the continuous development of technology (Baykal and Baykal, 2008: 10). In the process of industrialization and technological development, environmental pollution was addressed as a problem first in the Western European countries and then in all other countries around the world (Bozkurt, 2016: 8). However, the main reason for environmental degradation is not these phenomena and processes themselves, but the policies that paved the way for the occurrence of these phenomena and processes in an unplanned, poor and irregular way. Today, prioritization of the sense of consumption for production and the goal of economic growth in line with the conceptualization of welfare in societies is one of the most important factors that increase the human impact on the environment.

Although there are many reasons for environmental problems, it is possible to list some of them as follows: tourism, poverty, technological developments, industrialization, urbanization and population growth. We can also group the main reasons for environmental problems under three headings: industrialization, urbanization and population.

2.1.1 Industrialization

Agricultural and industrial revolutions are considered as turning points in the history of humanity. These two events caused great changes in the socio-cultural areas of social life (Turgut, 2012: 8). Industrialization is considered as a prerequisite for socio-economic development which is required for the creation of an advanced artificial environment (Ertürk, 2012: 125). Since the industrial revolution, the idea that the quality of human life can be improved through economic and physical processes and that social welfare depends on the achievement of economic goals has become widespread. In line with this idea, increasing production and consumption activities have resulted in the depletion of limited resources on the one hand and environmental problems on the other (Bozkurt, 2016: 14).

Industrialization leads to over-exploitation and rapid depletion of natural resources on the one hand, and the release of liquid, solid and gaseous wastes into the environment which causes pollution on the other. This disrupts the natural energy flow and material cycles in the ecosystem and increases the amount of wastes that do not decompose in the natural environment and cannot be recycled, leading to the pollution of the environment (Ertürk, 2012: 126). In order to prevent air pollution caused by industries, countries make amendments to their national legislation.

2.1.2 Urbanization

Urbanization is the term that refers to the process of population concentration which, in the strict sense, leads to an increase in the number of cities and in the population living in cities, and in the broad sense, leads to an increase in the number of cities and the growth of cities as a result of industrialization and economic development, as well as causing increased division of labor, organization and specialization in the social structure and urban-specific changes in human behaviors and relationships (Keleş, 2012: 31).

In 1927, Turkey had a total population of 13,648,270 people. Of these, 10,342,391 people were living in towns and villages and 3,305,879 people were living in provinces and districts. In percentage terms, 75.8 % of the population was living in towns and villages and 24.2 % were living in provinces and district (TURKSTAT, 2017). According to the population statistics of 2017 based on Address Based Population Registration System (ABPRS), Turkey has a total population of 80,810,525 people. Of these, 74,671,132 people live in provinces and districts and 6,049,393 people live in towns and villages. In percentage terms, 92.5 % of Turkey's population lives in provinces and districts, whereas 7.5 % live in towns and villages (TURKSTAT, 2017). In the last 90 years, the population living in towns and villages decreased by 68.3 %, from 75.8 % to 7.5 % in Turkey. On the other hand, the population living in provinces and district increased from 24.2 % to 92.5 %. In other words, according to the 2017 census data, 92 out of every 100 people live in urban areas in Turkey. With the Law No. 6360 published in 2012, the villages within the borders of the metropolitan cities were converted into neighborhoods. Therefore, the population living in the villages of metropolitan cities now lives in the districts of metropolitan cities. Although this specific situation affects the recent scenario to a certain extent, the overall picture shows that rural-to-urban migration increased rapidly in Turkey over the years and most of the people now live in city centers. The urbanization rate tends to increase around the world and 54 % of the world's population lives in urban areas according to United Nations (UN) 2014 data. This rate is expected to increase up to 66 % by 2050 (Bozkurt, 2016: 13). The rapid increase in urbanization may lead to some urban-specific social and environmental problems.

Although urbanization contributes to the economic and social development of societies, it has some negative impacts such as air, water and noise pollution and excessive use of land. Among the environmental problems caused by urbanization are the problem of disposing household and industrial wastes in city centers, housing problem, industrialization caused by intensive production, air pollution caused by traffic, infrastructural problems, unplanned urbanization, destruction of green spaces and damages to historic urban fabric (Bozkurt, 2016: 13).

2.1.3 Population

Rapid population growth is one of the main factors that cause an increase in environmental problems, especially environmental pollution (Ertürk, 2012: 131). Due to unplanned industrialization, urban sprawl and rapid population growth, the world's limited resources are under pressure.

Although it is not right to interpret the problem of population as a numerical increase that occurs in some countries, population growth is a problem area because it is an element of pressure on natural resources. The main reason why environmental policies address the population problem is the relationship between population and environment (Keleş et al., 2012: 105). The increase in all consumption needs and changing habits and expectations in response to population growth cause natural resources to remain inadequate as well as environmental degradation and ecological imbalance (Bozkurt, 2016: 11).

While a large portion of the earth remains empty due to climate and roughness, there is overpopulation in certain regions and 8 % of the world's surface area contains half of the population (Keleş et al., 2012: 105). 75 % of the world's population lives in underdeveloped and developing countries and 25 % lives in developed countries. However, it is emphasized that the root of the problem is related to production and consumption patterns, not to the numerical increase in population (Bozkurt, 2016: 11).

2.2 Global Environmental Problems

The variety and the number of environmental problems increase due to many reasons. Apart from the classical environmental problems, there are many other environmental problems ranging from global warming, loss of biodiversity and nuclear pollution to space pollution, housing problems, drought and nutrition problems. Most of these problems are global in terms of their effects and consequences.

Global environmental problems are problems of global scale experienced by most of the countries around the world to varying extents (Turgut, 2012: 7). Various human interventions on the ecosystem exceed the time and place dimensions, turning into global problems through the direct or indirect effects of factors other than humans (Kaplan, 1999: 16).

Today, environmental problems that occur in many countries and threaten the whole world can be classified under several headings, although their extent and impacts may change. These may be listed as follows: air-water-land pollution which are also referred to as classical environmental problems; climate change and global warming; ozone layer depletion; deforestation; loss of biodiversity;

depletion of natural resources; acid rain, erosion and desertification. In addition, it is possible to talk about various other environmental problems such as visual pollution, noise pollution, radioactive pollution, pollution from different kinds of wastes, electromagnetic pollution, destruction of wetlands, coastal pollution and destruction of cultural and natural assets.

3 Environmental Awareness

Following the end of the two world wars of the 20th century, the issue of environmentalism has been given a place in the common agenda together with the issues of peace, freedom and development. Environmentalism covers the activities carried out to protect, develop and improve the environment including protection of natural resources, prevention of environmental pollution, disposal of hazardous wastes, protection of historical and cultural assets and prevention of biodiversity loss through protection of plant and animal species (Alnıaçık, 2009: 51).

The potential of environmental problems to threaten the quality of life and health of communities requires finding solutions to these problems. Finding long-term solutions for environmental problems and ensuring the sustainability of these solutions is only possible through including all individuals and institutions in the process. Therefore, it is a need to increase environmental awareness and consciousness among all individuals around the world (Akçay and Pekel, 2017: 1176).

There are various campaigns run worldwide on issues such as environmental awareness, green concern, reduction of consumption, change of harmful and wasteful consumption patterns, waste collection and preference toward using green eco-friendly products. Such campaigns affect people's consumption behaviors (Alkaya, Çoban, Tehci and Ersoy, 2016: 122).

All sectors are expected to promote environmental awareness and all companies are expected to fulfill their responsibilities toward the environment during their operation. Examining the health care sector, which is an important employment area in Turkey with a share of 5 % in GDP on a yearly basis, in the context of environment will allow the other sectors to make self-assessment in this regard.

4 Health and Healthcare Sector

Health refers to a state of complete physical, mental and social well-being and not merely the absence of disease or infirmity, and the healthcare sector is a general and comprehensive term used to describe the systems and subsystems (as well as persons, institutions, organizations, status, products etc. that these systems and subsystems include) established in different areas of production to produce/supply and demand/consume all kinds of goods and services that have direct or indirect or principal effects on health (Sargutan, 2005). If the institutions and organizations that provide healthcare services adopt sustainable policies in terms of economic, social and environmental factors and they implement these policies, this will help them and the sector in which they operate develop cumulatively. This is because development is a multidimensional concept that is realized collectively. Therefore, for the development of the sector, it is important that all institutions and organizations operating in the field of health sector from pharmaceutical and medical device companies to hospitals should establish systems with control mechanisms and use them during their operation without compromising the sustainability principles.

In Turkey, the share of healthcare sector in Gross Domestic Product (GDP) was around 4.6 % in 2016. This rate is below the OECD average of 8.9 % and can be said to remain at a sustainable level for years. Although the number of institutions and organizations operating in the healthcare sector increase every year according to annual health statistics, a total of 1510 hospitals, 876 of which are affiliated to the Ministry of Health, 69 of which are under universities and 565 of which are private, provide healthcare services as of 2016. According to 2016 statistics, there were 871,334 people working in all healthcare institutions and organizations under the Ministry of Health, universities and private sector. It is expected that institutions in which nearly one million people are employed, from which 648,381,615 patients received services in the last five years and which produce 81 thousand tons of medical waste are of green and environmentally friendly nature and are built using eco-friendly technologies and designs. Environmentally-friendly buildings that employ technologies that do not harm the environment, that have the infrastructure allowing the efficient and effective use of resources and energy, that use renewable energy sources or generate their own energy and that recycle wastes or make them harmless (Şenocak and Bursalı, 2018), will make significant contributions to the sustainable healthcare sector.

5 Environmentally-Friendly Buildings

Buildings are assessed and certified to have green properties based on the established and standardized criteria. Anbarci, Giran and Demir (2012) list the common certificate systems as LEED (Leadership in Energy Efficiency Design), BREEAM

(Building Research Establishment Environmental Assessment Method), DGNB (Deutsche Gesellschaft für Nachhaltiges Bauen), IISBE (International Initiative for Sustainable Built Environment), Greenstar (Environmental Rating System for Buildings), CASBEE (Comprehensive Assessment System for Built Environment Efficiency).

The institutions in the healthcare sector that first come to mind are hospitals and healthcare institutions and organizations that provide healthcare services. However, companies and laboratories that produce medical consumables, medical devices, pharmaceutical medical products and non-classified by-products required during the provision of healthcare services are important indicators that the healthcare sector is not only services sector, but also a manufacture-based sector. Facilities operating in this sector have some impacts on the environment during their activities. These effects may be caused by solid, liquid and gaseous wastes, as well as by the use and consumption of energy sources in the category of all raw materials that produce energy required for the execution of activities.

6 Healthcare Sector: Key Features, Waste Management and Sustainability

Healthcare services have positive externalities. The reason is because those who cannot reach healthcare services are affected indirectly by the same services and their consequences as much as those who can reach these services. For example, as part of preventive health care, vaccination of a person protects others from being infected, too. Therefore, healthcare services are non-profit social services (Çakmak, Öktem and Ömürgönülşen, 2009). Healthcare is a service that is produced in the market because of its special benefits at national (local) level and its excludable feature as well as the competitiveness in the consumption of these services. However, its two important externalities which have both local and transboundary impacts cause healthcare to have the characteristics of a public good. The first is the global nature of infectious diseases; an infectious disease is dangerous as a public bad that affects other people and countries. The second externality is that external benefits created by protecting a person or country (as in the case of bird flu) from infectious diseases also reduce the risk for other people and countries (Mutlu, 2006). Some specific characteristics of the healthcare sector can be effective in the emergence of environmental problems caused by the healthcare sector.

Health care facilities produce wastes as a result of the use of consumables that might be infectious or might lead to pathological, perforating injuries as well as wastes as a result of using medicines, vaccines, serums and other pharmaceutical substances; wastes of different physical properties produced from chemical substances; wastes that include heavy metals such as mercury, cadmium and lead or that have the characteristics of heavy metals; bodily fluids such as blood and blood products; human tissues removed in the course of surgical interventions; consumables contaminated with these products; air filters used to retain bacteria; microorganism cultures used in laboratories and a number of waste materials during the during the medical procedures. The collection of wastes by sorting them into categories based on their types is important in terms of using the disposal method appropriate to the characteristics of each type. All waste generated in hospitals carries a risk for public health and must be destroyed or recycled using appropriate waste management procedures.

Pathogens, concentrated cultures and contaminated needles (especially hypodermic needles) present a potential health risk. Needles are dangerous because they carry the risk of being contaminated not only by injection, but also by pathogenic agents. Pathogens in infected wastes can be transmitted to the host through the skin in cases where skin integrity is impaired, such as fresh or open wounds. Failure to reduce sufficiently the generation of waste at its source, the lack of public awareness of recycling, inadequate waste disposal systems in rural areas, the insufficient monitoring of movement of hazardous waste, insufficient waste disposal capacity and inadequate inspection of the implementation of the legislation (İlter, 2014) are among the main waste management problems in Turkey.

The Regulation on the Control of Medical Waste (2017) outlines a number of principles regarding the management of medical wastes and the liabilities of healthcare institutions. As per this regulation, healthcare facilities that produce waste are not allowed to establish and operate individual medical waste treatment facilities. Besides, they are responsible for periodically training the staff in charge of medical waste management activities in municipalities or in companies with environmental license furnished with authority by municipalities which are responsible for collecting, carrying and disposing wastes, and for ensuring that the staff goes through medical examinations periodically. They are also responsible for ensuring that the activities within the scope of medical waste management are carried out by these staff.

The Green hospital concept has been discussed in the context of environmental awareness in Turkey recently. Green hospital refers to producing alternative resources, promoting more efficient and effective use of energy, water and materials, preventing all kinds of wasting and adopting a plain management approach, developing environmentally friendly and sensitive building designs and being environmentally friendly during the provision of services (Özkan, Bayın and Yeşilaydın,

2014). In recent years, studies have been conducted to examine whether healthcare facilities in Turkey are environmentally conscious in their daily operations (Toker and Çınar, 2017; Özkan et al., 2014; Palteki, 2013; Çilhoroz and Işık, 2018).

Toker and Çınar (2017) examined the sustainability structures of the hospitals in Istanbul and found out that there is no significant difference between the ownership structure of hospitals and sustainable environmental policy implementations. They also reported that sustainable environmental policy implementations of both public and private hospitals are not at the desired level and sustainable social policies are at a higher level of implementation in both public and private hospitals.

Palteki (2013) made an assessment of the public hospitals in Istanbul, which provide a significant amount of healthcare services in Turkey, in terms of the sustainability of healthcare services and greenness by using the "Green Hospital Qualification Assessment Form" developed by using the standard forms. She assessed the compliance of the hospitals to meet the green assessment criteria for hospitals and found that the hospitals examined in the study scored the highest in environmental management systems, showing 91.4 % conformity with the green hospital criterion; and scored lowest in water management, with a score of 45.4 %. In the study, the average score for the hospitals' conformity with the green hospital criteria was found to be 68.6 %.

Çilhoroz and Işık (2018) examined the conformity of hospitals in Ankara with the green hospital criteria and found that the general conformity of public hospitals, private and all hospitals in Ankara with the green hospital standards were 71.8 %, 72.5 and 72.2 respectively. They also compared the LEED for healthcare standards with the conformity of hospitals with green hospital standards and reported that both public and private hospitals met these standards in areas of water management and sustainable facilities and they did not meet the standards in areas of energy management and materials selection.

In Turkey, there are some healthcare institutions and organizations certified using green building rating systems. Furthermore, planning of newly constructed city hospitals by taking account of the green hospital standards can be listed as one of the positive developments toward sustainable health institutions. In addition, it would be appropriate to ensure sustainability at the local level through good practice examples in the health sector.

Conclusion

It is seen that the green hospital concept has been addressed recently within the scope of environmental consciousness of the healthcare sector in Turkey. It will contribute to promoting the awareness of sustainable environment, if institutions and organizations operating in the healthcare sector, especially the healthcare facilities, prepare a proper waste plan and segregate waste at the source, carry out in-house handling of wastes, temporarily store them, transport them to disposal areas and finally dispose them, treat recyclable waste separately and use all kinds of energy efficiently.

Despite the comprehensive literature analyzing environmental issues, there is no objective model that can be used to assess the environmental sustainability of health organizations, to compare the results obtained for an organization in a given time and to establish a comparison or benchmarking tool to analyze difference between institutions. Therefore, further studies may propose a model. It would be appropriate if more controlled physical environments are created in hospitals by adopting the understanding that hygiene constitutes three quarters of health.

References

Akçay, S. ve Pekel, F. O. (2017) "Öğretmen Adaylarının Çevre Bilinci ve Çevresel Duyarlılıklarının Çeşitli Değişkenler Açısından İncelenmesi", Elementary Education Online, 16(3): 1174–1184.

Akdur, R. (2005) "Avrupa Birliği ve Türkiye'de Çevre Koruma Politikaları: Türkiye'nin Avrupa Birliğine Uyumu", Ankara: Ankara Üniversitesi Avrupa Topluluğu Araştırma ve Uygulama Merkezi Araştırma Dizisi, No: 23.

Alkaya, A., Çoban, S., Tehci, A. ve Ersoy, Y. (2016) "Çevresel Duyarlılığın Yeşil Ürün Satın Alma Davranışına Etkisi: Ordu Üniversitesi Örneği", Erciyes Üniversitesi İktisadi ve İdari Bilimler Fakültesi Dergisi, 47: 121–134.

Alnıaçık, Ü. (2009) "Tüketicilerin Çevreye Duyarlılığı ve Reklamlardaki Çevreci İddialar", Kocaeli Üniversitesi Sosyal Bilimler Enstitüsü Dergisi, 18(2): 48–79.

Anbarcı, M., Giran, Ö. ve Demir, İ. H. (2012) "Uluslararası Yeşil Bina Sertifika Sistemleri ile Türkiye'deki Bina Enerji Verimliliği Uygulaması", E-Journal of New World Sciences Academy, 7(1): 368–383.

Baykal, H. ve Baykal, T. (2008) "Küreselleşen Dünya'da Çevre Sorunları", Mustafa Kemal Üniversitesi Sosyal Bilimler Enstitüsü Dergisi, 5(9): 1–17.

Bozkurt, Y. (2016) "Çevre Sorunları ve Politikaları", Bursa: Ekin Yayınevi.

Çakmak, M., Öktem, M. K. ve Ömürgönülşen, U. (2009) "Türk Kamu Hastanelerinde Teknik Verimlilik Sorunu: Veri Zarflama Analizi Tekniği ile Sağlık Bakanlığı'na Bağlı Kadın Doğum Hastanelerinin Teknik Verimliliklerinin Ölçülmesi", Hacettepe Sağlık İdaresi Dergisi, 12(1): 1–36.

Çilhoroz, Y. ve Işık, O. (2018) "Ankara'daki Hastanelerin Yeşil Hastane Ölçütlerine Uygunluğunun İncelenmesi", Hacettepe Sağlık İdaresi Dergisi, 21(1): 41–63.

Ertürk, H. (2012) "Çevre Bilimleri", Bursa: Ekin Yayınevi.

Görmez, K. (2010) "Çevre Sorunları", Ankara: Nobel Yayın Dağıtım.

İlter, H. (2014) "Atık Yönetimine Sağlık Bakanlığının Yaklaşımı", II. Ulusal Sağlık Kuruluşları Çevre Yönetim Sempozyumu Kitabı, İstanbul, 24–30.

Kaplan, A. (1999) "Küresel Çevre Sorunları ve Politikaları", Ankara: Mülkiyeliler Birliği Vakfı Yayınları.

Karacan, A. (2007) "Çevre Ekonomisi ve Politikası", İzmir: Ege Üniversitesi İktisadi ve İdari Bilimler Fakültesi Yayını.

Keleş, R., Hamamcı, C. ve Çoban, A. (2012) "Çevre Politikası", Ankara: İmge Kitabevi.

Mutlu, A. (2006) "Küresel Kamusal Mallar Bağlamında Sağlık Hizmetleri ve Çevre Kirlenmesi: Üretim, Finansman ve Yönetim Sorunları", Maliye Dergisi, 150: 53–78.

Özkan, O., Bayın, G. ve Terekli Yeşilaydın, G. (2014) "Hastane Yönetiminde Sürdürülebilir Yaklaşım: Yeşil Yönetim", 8. Sağlık ve Hastane İdaresi Kongresi, Lefke- Kıbrıs, Proceeding Book, 2238–2248.

Palteki, A. S. (2013) "İstanbul'daki Kamu Hastanelerinin Yeşil Hastane Ölçütlerine Uygunluklarının Belirlenmesi", İstanbul: Yüksek Lisans Tezi, İstanbul Üniversitesi, Sağlık Bilimleri Enstitüsü.

Resmî, G. (1983) "2872 Sayılı Çevre Kanunu", 18132.

Sağlık, B. (2017) "Sağlık İstatistikleri Yıllığı 2016", https://dosyasb.saglik.gov.tr/Eklenti/13183,sy2016turkcepdf.pdf?0, (01.08.2018).

Sargutan, E. (2005) "Sağlık Sektörü ve Sağlık Sitemlerinin Yapısı", Hacettepe Sağlık İdaresi Dergisi, 8(3): 400–428.

Şenocak, B. ve Mohan Bursalı, Y. (2018) "İşletmelerde Çevresel Sürdürülebilirlik Bilinci ve Yeşil İşletmecilik Uygulamaları ile İşletme Başarısı Arasındaki İlişki", Süleyman Demirel Üniversitesi İktisadi ve İdari Bilimler Fakültesi Dergisi, 23(1): 161–183.

Sipahi, E. B. (2010) "Küresel Çevre Sorunlarına Kolektif Çözüm Arayışları ve Yönetişim", Selçuk Üniversitesi Sosyal Bilimler Enstitüsü Dergisi, 24: 331–344.

Toker, K. ve Çınar, F. (2017) "Sağlık Sektöründe Kurumsal Sürdürülebilirlik Yönetimi ve İstanbul Avrupa Yakasında Faaliyette Bulunan Hastanelerde Bir Araştırma", International Conference on Eurasian Economies, Session 4C: Sektörel Analizler, 412–417.

TÜİK. (2017) "Yıllara Ve Cinsiyete Göre İl/İlçe Merkezleri Ve Belde/Köyler Nüfusu: 1927-2017", http://www.tuik.gov.tr/UstMenu.do?metod=temelist, (30.07.2018).

Turgut, N. (2012) "Çevre Politikası ve Hukuku", Ankara: İmaj Yayınevi.

UN. (1997) "Glossary of Environment Statistics", Series F. No. 67. New York: United Nations Publication.

Yalçın, A. Z. (2009) "Küresel Çevre Politikalarının Küresel Kamusal Mallar Perspektifinden Değerlendirilmesi", Balıkesir Üniversitesi Sosyal Bilimler Enstitüsü Dergisi, 12(21): 288-309.

Mustafa Cem Aldağ

Impacts of Technological Developments on the Environment and Agriculture

Introduction

The agricultural sector is intertwined with the nature. Natural factors affect productivity directly. As a result, environmental pollution and changes in compositions of natural resources such as soil and water that are highly important for agriculture affect the quality and amount of agricultural products negatively. As animal production and herbal production constitute an organic whole, the most important protein or food sources of humans are directly or indirectly affected by environmental pollution. As the economic development rate increases, the damage of countries on the environment increases as well (Karaer and Gürlük 2003).

Presence of a balanced correlation between agriculture and environment is explained with the "sustainable agriculture" concept. Sustainable agriculture requires management of natural resources in a way that they will be useful in the future. There must be a balance between using land and natural resources in a useful manner and environmental protection.

Agricultural activities that have been carried out for hundreds of years are the main way of life for a vast majority of the world. In years, agricultural activities have gradually developed and become more efficient. Furthermore, new research and technological advancements reveal negative impacts of agriculture on the environment.

Although agriculture is in a complicated relationship with natural resources and the environment, it is hard to say that certain environmental impacts are caused by agriculture and this has not been completely understood yet. Agriculture is the main user of land and water sources and maintaining the amount and quality of these resources is necessary for sustainability of agriculture. Agriculture both creates waste and pollution and protects natural resources and contribute to their cycle. Furthermore, it changes the surrounding countryside and living areas for wildlife. Many environmental impacts remain within the sector but impacts of agriculture outside land are also important. Although some of them are nationally and internationally important, they are mostly local and regional.

It is generally agreed that environmental performance of agriculture must be increased by increasing its useful impacts and reducing the harmful impacts on the environment in order to maintain continuous use of resources.

As for global climate change, some of the most important problems are related to the impacts of climate change on agriculture, forests, water sources etc. and whether they could adapt to the changing climate. Although attention has been drawn to problems related to climate change faced by the world for a long time, precautions have been taken to reduce emission of greenhouse gases with the aim of preventing climate change from reaching dangerous levels and some steps have been taken in connection with obligations of countries regarding this matter, we cannot say that these measures are enough.

1 Environmental Impacts of Agriculture

The impacts of agricultural activities on water sources, the negative impacts of water taken from irrigation canals for irrigation purposes on the water resource depending on the structure of the water source and pollution of water sources resulting from contamination of water sources by agricultural fertilizers and pesticides can be given as examples of this.

Irrigation during intense agricultural activities prevents sustainable use of water. Land drainage can decrease the underground water level and destroy wetlands where many species live. Furthermore, pesticides and nitrate produced by fertilizers can leak into underground water while nitrogen and phosphorus produced by fertilizers can mix with surface flows. Irrigation and drainage can affect underground water levels and cause soil sodification. Accumulations caused by erosion can have negative impacts on water sources and wetland ecosystems.

The amount of water used for irrigation can cause changes in climate, product range, soil characteristics, water quality, soil cultivation applications and irrigation methods. Productivity is increased and risks in dry periods are decreased with irrigation, thus making it possible to grow more profitable crops.

Efforts have been initiated in many countries for treatment and reuse of wastewater in order to meet the demand for water sources. The USA, Israel, Western Europe and Australia are pioneers in this field. Treatment and reuse of wastewater has become an effective method of saving water and meeting the increased water need. When treatment is considered for reuse of water, it is questioned whether it is possible to use the water for this purpose. High levels of treatment and reliability are needed especially in irrigation areas and for growing fruits and vegetables while the same sensitivity is not shown while using water for other purposes. Some standards must be determined for discharge point and intended use of waste water. These standards can be chemical, biological and physical parameters, and the irrigation method and vegetation pattern related to reuse of the waste water must be taken into account while determining them (Yurtseven et al. 2010).

Excessive use of water, outdated irrigation networks, water pollution, performing water delivery and distribution with open systems, and organization and management problems can be listed as problems encountered in connection with the irrigation method (Çakmak, Yıldırım and Aküzüm 2008).

Another result of over-irrigation is the need for building larger distribution and drainage networks, using more resources and using additional energy if there is a pump in the system. At this point, people have started to produce environmental solutions in order to power pumps and it is predicted that pumps will be used more commonly when they become more affordable.

Since the increase in price per increase in unit power output of a photovoltaic system is greater than that for a diesel, gasoline or electric system, photovoltaic power is more cost competitive when the irrigation system with which it operates has a low total dynamic head. For this reason, photovoltaic power is more cost-competitive when used to power a micro irrigation system as compared to an overhead sprinkler system. Photovoltaic power for irrigation is cost-competitive with traditional energy sources for small, remote applications, if the total system design and utilization timing is carefully considered and organized to use the solar energy as efficiently as possible. In the future, when the prices of fossil fuels rise and the economic advantages of mass production reduce the peak watt cost of the photovoltaic cell, photovoltaic power will become more cost-competitive and more common (Eker 2005).

The necessity to increase agricultural production in parallel with the increase in the world population causes more inputs to be used and risks posed by some artificial fertilizers that are the most important inputs create a dilemma. High productivity through plant nutrition, requirements for high quality and healthy products, determining the fertilizer need, types and amounts of fertilizers, application method, application frequency and time, application records, fertilizer storage, records related to organic fertilizers etc. are important. Especially new applications emerging in parallel with developments in information level and technology indicate that healthy and high quality agricultural production expected by the modern society can be realized with conscious producers. Training producers that do not damage the environmental and human health, protect natural resources, ensure food safety, perform agricultural production whose all stages can be monitored must be regarded as the biggest social service. The fact that contemporary quality management systems such as Good Agricultural Practices have started to be used in Turkey and a lot of progress has been made in a short period of time gives hope for the future (Karaçal and Tüfenkçi 2010).

According to the research, the contamination of the environment with micro plastic, defined as particles smaller than 5 mm, has emerged as a global challenge because it may pose risks to biota and public health. Current research focuses predominantly on aquatic systems, whereas comparatively little is known regarding the sources, pathways and possible accumulation of plastic particles in terrestrial ecosystems. Particles were classified by size and identified by attenuated total reflection-Fourier transform infrared spectroscopy. All fertilizer samples from plants converting bio-waste contained plastic particles, but amounts differed significantly with substrate pretreatment, plant and waste type. In contrast, digesters from agricultural energy crop digesters tested for comparison contained only isolated particles, if any. The results of the research indicate that depending on pretreatment, organic fertilizers from bio-waste fermentation and composting, as applied in agriculture and gardening worldwide, are a neglected source of micro plastic in the environment (Weithmann et al. 2018).

Agricultural protection refers to protecting plants from impacts of diseases and weeds in an economical manner and increasing product amount and quality. As you can understand from this simple definition, agricultural protection aims to increase product amount and quality as well as being economical. In order to achieve this goal, agricultural protection must be carried out in accordance with integrated protection (integrated pest management). Integrated pest management refers to activities that benefit from all known methods in agricultural protection and have the minimum impacts on human and environmental health (Delen et al. 2005).

Different types of pesticides can have different impacts on the same organism. Pesticides can penetrate fatty tissues of organisms and accumulate in the food chain in an ascending order. Human and animal health can be negatively affected through direct contact.

Most insecticides used today are neurotoxic and reveal their effects by affecting the neural system of target organisms. Some studies indicate that insecticides affect the reproductive system, blood production, liver, endocrine system, urinary system and other systems in addition to the neural system in vertebrates. It has been reported that pesticides containing organochlorine cause irregularities in renal tissues, glomeruli and tubules of advanced vertebrates, undulations in glomerular basal membrane, deformation in general appearance of tubules caused by vacuolar degeneration and mononuclear cell infiltration as well as hydropic degeneration (Önen, Fındık Güvendi and Beşeren 2016).

Pesticide consumption in Turkey increased by 45.29 % in 2002 compared to 1979. Despite this increase, pesticide consumption in Turkey is considerably low compared to developed countries. However, consumption in regions such as the Mediterranean Region and Aegean Region where intensive agriculture is performed is much higher than the Turkey average. Although the general pesticide consumption is low in Turkey, the most consumed pesticides pose environmental and health risks. (Delen et al. 2005)

There is no doubt that global warming will have significant impacts on water supply and changes in rainfall variability will create serious problems in the agricultural sector (Hoffman and Evan 2007).

Carbon dioxide emissions resulting from agricultural activities increased as of the mid-1880s and agriculture became the main source of carbon dioxide emissions before the 1920s. Today, the main source of carbon dioxide is use of fossil fuels while the second most important source of increased carbon dioxide emissions is land conversions carried out for agriculture. Converting uncultivated lands into farm lands causes carbon dioxide emission into the air as the vegetation cover is removed. Agricultural activities account for 20 % of carbon dioxide emission caused by all human activities in the world (Rohila et al. 2017).

The agricultural sector is the driving force behind gas emission and it is thought that land usage contributes to climate change. Agriculture is an important user of land and fossil fuels and practices such as rice cultivation and livestock raising make a direct contribution to emissions of greenhouse gases. It was stated at the Intergovernmental Panel on Climate Change that the increase in greenhouse gases in the last 250 years had three main reasons: fossil fuels, land use and agriculture (Anonymous 2018).

Agriculture contributes to the increase in greenhouse gases through land use mainly in four ways:

- CO_2 emissions resulting from deforestation,
- Methane emissions resulting from rice cultivation,
- Methane emissions resulting from enteric fermentations of bovine animals,
- Nitrogen oxide emissions resulting from fertilizer use.

These agricultural practices constitute 54 % of methane emissions, approximately 80 % of nitrogen oxide emissions and all carbon dioxide emissions resulting from land use (Prather et al. 1995).

Tab. 1: Annual Greenhouse Gas Emissions by Sector, in 2010. Source: Edenhofer et al. 2014

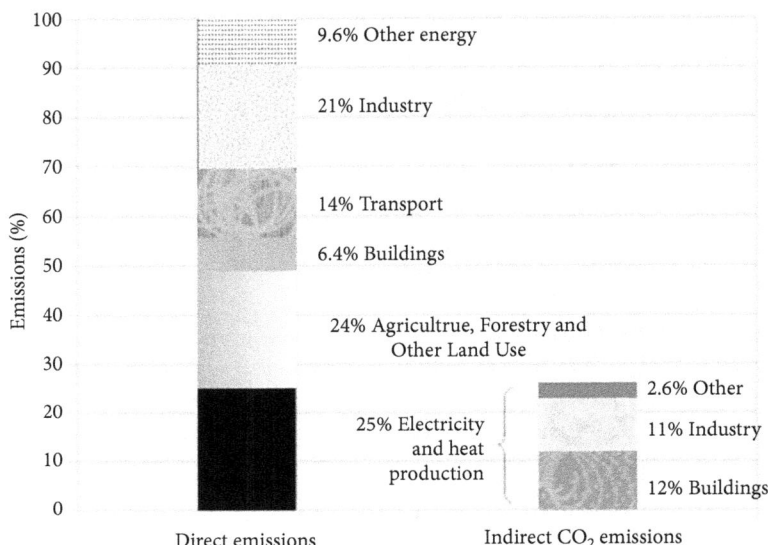

Livestock raising, deforestation and activities related to livestock raising making intensive use of fuel account for more than 18 % of human-induced greenhouse gas emissions including the emissions given below.

- Carbon dioxide emissions on a global scale: 9 %,
- Methane emissions on a global scale: 35–40 % (mainly resulting from enteric fermentation and animal manure),
- Nitrogen oxide emissions on a global scale: 64 % (mainly resulting from fertilizer use (Steinfeld et al. 2006).

Anaerobic decomposition is formation of carbon dioxide (CO_2), methane (CH_4) and a little hydrogen sulphide (H_2S) in airless (oxygen free) environments. The gas mixture that consists of 60–70 % methane (CH_4), 30–40 % carbon dioxide (CO_2) and 0–2 % hydrogen sulphide (H_2S) resulting from decomposition of organic substances in an airless environment is called biogas (Yılmaz et al. 2017). In other words, renewable energy can be generated by burning animal manure.

Biogas technology makes it possible to both generate energy from waste materials and recycle waste into the soil. Some of its benefits are listed below.

- Biogas is a source of cheap and environmentally friendly energy and fertilizer.

Tab. 2: Components of Biogas. Source: Angelidaki, LyhneLuo and Gang 2014

Compound	Formula	%
Methane	CH_4	50–75
Carbone dioxide	CO_2	25–50
Nitrogen	N_2	0–10
Hydrogen	H_2	0–1
Hydrogen Sulfide	H_2S	0–3
Oxygen	O_2	0–0

- Enables waste recycling.
- Smells of animal manure fades away and becomes undetectable thanks to biogas generation.
- Diseases that are caused by animal manure and threaten human health and underground water lose their effect to a great extent.
- Waste does not disappear after biogas generation, but they become much more valuable organic fertilizers.
- Biogas is a clean source of energy with high heating value.
- Weed seeds in animal manure lose their ability to germinate due to biogas generation (Yılmaz et al. 2017).

Desertification, erosion, reduction in organic substances in soil, soil pollution, soil compaction, soil biodiversity decrease and salinity can reduce the capacity of soil to fulfill its main functions. Such decomposition can be caused by incorrect agricultural activities such as unbalanced use of fertilizers, extracting excessive amounts of underground water for irrigation, over-irrigation, insufficient land drainage, incorrect use of pesticides, use of heavy machinery or equipment and overgrazing etc. Abandoning certain agricultural practices can contribute to decomposition of soil.

Fertilizer use can impact soil reaction, structure, edaphon and accumulation of toxic substances in soil. However, chemical fertilizers effect basic qualities of soil only after they are used for a very long time, unilaterally and in the same form every year and such impacts are not very severe and negative (Taşkaya 2004).

Genetic improvement of crops has proven to advance yields over the past 50 years and as (Duvick 2005) pointed out, yield increases in maize can be attributed equally between genetic improvement and management improvements (Hatfield and Walthall 2015).

(George 2014) reports that more attention must be paid to development of agricultural practices despite the advancements in germplasm and agriculture.

Increasing productivity is considered in parallel with increasing incomes of producers in developing countries. (Mueller et al. 2012) point out that productivity gap can be closed by a combination of intensive nutrient use and water management. However, they did not acknowledge the presence of climate change and they did not take into account possible production setbacks that prevent achieving the potential productivity due to the climate change.

Soil biodiversity plays a key role in regulating the processes that underpin the delivery of ecosystem goods and services in terrestrial ecosystems. Agricultural intensification is known to change the diversity of individual groups of soil biota, but less is known about how intensification affects biodiversity of the soil food web as a whole, and whether or not these effects may be generalized across regions (Tsiafouli et al. 2015).

An important goal of ecological compensation measures in farm lands is to maintain and improve regional product diversity. On the other hand, some available European agricultural and environmental programs seem highly ineffective. A possible explanation for this could be the lack of source populations in planted lands. Residues of natural and semi-natural habitats can make various contributions to regional biodiversity (Duelli and Obrist 2003).

Many authors and organizations worldwide give their own definition of sustainable agriculture. Overall, all authors agree on the occurrence of three approaches to the concept of sustainable agriculture: environmental, economic and social approaches. In other words, agricultural systems are considered to be sustainable if they sustain themselves over a long period of time, that is, if they are economically viable, environmentally safe and socially fair. Beyond this ideological definition, the practical issue is to build operational solutions to reach global goals. This is a challenging task because the stakeholders do not agree on the criteria to measure the sustainability of a farming system, and on how to balance those criteria. Many indicators have indeed already been produced to evaluate sustainability. (Lichtfouse et al. 2009).

Sustainable agriculture refers to an agricultural structure in which natural resources are protected in the long term and agricultural technologies that do not harm the environment are used (Turhan 2005).

The three main objectives of sustainable agriculture are: environmental health, economic profitability and social and economic equality.

Today, sustainable farming practices commonly include:

- Crop rotations that mitigate weeds, disease, insect and other pest problems; provide alternative sources of soil nitrogen; reduce soil erosion; reduce risk of water contamination by agricultural chemicals,

Tab. 3: Most Common Indicators of Sustainable Agriculture Themes among Existing Indices, Reports and Datasets Reviewed. Source: Reytar, Hanson and Henninger 2014

Indicator theme	Indicator (example)
Water use	Total water use for agriculture production
Agricultural policy related to government support	Agricultural subsidies
Climate change	Greenhouse gas emissions from agricultural sources
Agricultural production	Crop yield
Agricultural inputs	Fertilizer use
Land use	Area of agricultural land
Environmental policy	Participation in UNFCCC treaties
Environmental degradation	Area of degraded/barren lands
Ecosystem biodiversity	Wild species in agricultural lands
Water quality	Number of dead (hypoxic) zones
Agricultural Research and Development	Public agricultural research expenditures
Ecosystem management	Area of terrestrial reserves
Agricultural policy related to the environment	Pesticide regulations

- Pest control strategies that are not harmful to natural systems, farmers, their neighbors or consumers. This includes integrated pest management techniques, that reduce the needs for pesticides by practices such as scouting, use of resistant cultivars, timing of planting and biological pest controls,
- Increased mechanical /biological weed control; more soil and water conservation practices and strategic use of animal and green manures,
- Use of natural or synthetic inputs in a way that poses no significant hazard to man, animals or the environment. (O'connell 1992)

(Meynard et al. 2006) identified four different ways to design innovative agricultural systems for sustainable development:

- Inventing new farming systems, breaking off with the current ones,
- Identifying and improving farming systems built by the local stakeholders,
- Giving tools and methods to stakeholders to improve their own systems or evaluate those proposed by scientists,
- Identifying the economic, social and organization conditions that may help the actors to adopt alternative farming systems.

Conclusion

It is not possible to consider agriculture separately from the nature. Natural conditions have a direct impact on productivity that is the most important output of agriculture. Environmental pollution and changes in concentrations of natural resources such as water and soil that are highly important for agricultural activities affect qualities and amounts of agricultural products negatively. As plant production and animal production constitute an organic whole, the most basic and important nutritional sources of humans are directly or indirectly affected by environmental changes.

Considering that the world population is expected to exceed 9 billion by 2050, it can be argued that environmental damages can increase due to the pressures for more production per unit area. Results obtained in a lot of studies show that we are at a point where it is not possible to sustain agriculture with current intensive agricultural production methods.

As a requirement of monoculture agriculture that has been performed for generations, pesticides intended for agricultural protection are used in an uncontrolled manner, this threatens the environment and other species as well as jeopardizing human health and it is hard to keep under control as it increases resistance to pesticides.

In conclusion, today sustainable agriculture is highly important in order to create a society that satisfies its needs without jeopardizing future generations. Protection of water and soil sources, protection of natural resources, ensuring biodiversity, integrated pesticide management, increasing productivity of farm lands using methods suitable for agricultural production and organic agriculture can provide a significant solution for a sustainable life and agriculture.

References

Angelidaki, I., LyhneLuo, P., and Gang, L. (2014). US 2014/0342426A1. United States: United States Patent and Trademark Office. http://appft.uspto.gov/netacgi/nph-Parser?Sect1=PTO2&Sect2=HITOFF&p=1&u=%2Fnetahtml%2FPTO%2Fsearch-bool.html&r=2&f=G&l=50&co1=AND&d=PG01&s1=Angelidaki.IN.&OS=IN/Angelidaki&RS=IN/Angelidaki Access Date: 07.07.2018

Anonymous. (2018). Climate change and agriculture, http://en.wikipedia.org/wiki/Climate_change_and_agriculture Access Date: 07.07.2018

Çakmak, B., Yıldırım, M., and Aküzüm, T. (2008) "Türkiye'de Tarımsal Sulama Yönetimi, Sorunlar ve Çözüm Önerileri", *TMMOB 2. Su Politikaları*

Kongresi, 215–224. http://www.imo.org.tr/resimler/ekutuphane/pdf/10929.pdf Access Date: 10.07.2018

Delen, N., Durmuşoğlu, E., Güncan, A., Güngör, N., Turgut, C., and Burçak, A. (2005) "Türkiye'de Pestisit Kullanımı, Kalıntı ve Organizmalarda Duyarlılık Azalışı Sorunları", *Türkiye Ziraat Mühendisliği VI. Teknik Kongresi*. https://www.researchgate.net/publication/269064638 Access Date: 18.05.2018

Duelli, P., and Obrist, M. K. (2003) "Regional biodiversity in an agricultural landscape: the contribution of seminatural habitat islands", *Basic and Applied Ecology*, 4(2): 129–138. https://doi.org/10.1078/1439-1791-00140 Access Date: 11.07.2018

Duvick, D. N. (2005) "The contribution of breeding to yield advances in maize (Zea mays L.)", *Advances in Agronomy*, 86: 83–145.

Edenhofer, O., Pichs-Madruga, R., Sokona, Y., Farahani, E., Kadner, S., Seyboth, K., ... Minx, J. C. (2014). IPCC, 2014: Climate change 2014: mitigation of climate change. Contribution of working group III to the fifth assessment report of the intergovernmental panel on climate change. Cambridge, United Kingdom and New York, NY, USA. https://www.ipcc.ch/site/assets/uploads/2018/02/ipcc_wg3_ar5_full.pdf Access Date: 02.07.2018

Eker, B. (2005) "Solar powered water pumping systems", *Trakia Journal of Sciences*, 3(7): 7–11. http://www.abyaran.com/pdf/technical-papers/pumps/SOLAR powered water pumping systems.pdf Access Date: 18.05.2018

George, T. (2014) "Why crop yields in developing countries have not kept pace with advances in agronomy", *Global Food Security*, 3(1): 49–58. https://doi.org/10.1016/j.gfs.2013.10.002 Access Date: 21.12.2018

Glenn J. Hoffman, and Robert G. Evan. (2007) "In design and operation of farm irrigation systems", 2nd Edition, 1–32 St. Joseph, MI: American Society of Agricultural and Biological Engineers. https://doi.org/10.13031/2013.23684 Access Date: 14.07.2018

Hatfield, J. L., and Walthall, C. L. (2015) "Meeting global food needs: Realizing the potential via genetics × environment × management interactions", *Agronomy Journal*, 107(4): 1215–1226. https://doi.org/10.2134/agronj15.0076 Access Date: 11.07.2018

Karaçal, İ., and Tüfenkçi, Ş. (2010). "Bitki Beslemede Yeni Yaklaşımlar ve Gübre - Çevre İlişkisi". VII. Teknik Kongre, 257–268. http://www.zmo.org.tr/resimler/ekler/fc64354454711c9_ek.pdf Access Date: 18.05.2018

Karaer, F., and Gürlük, S. (2003) "Gelişmekte Olan Ülkelerde Tarım-Çevre-Ekonomi Etkileşimi", *Doğuş Üniversitesi Dergisi*, 4(2): 197–206. http://journal.dogus.edu.tr/index.php/duj/article/view/175 Access Date: 18.05.2018

Lichtfouse, E., Navarrete, M., Debaeke, P., Ere, V., Alberola, C., and Ménassieu, J. (2009) "Agronomy for sustainable agriculture", A review. *Agron. Sustain. Dev*, 29: 1-6. https://doi.org/10.1051/agro:2008054 Access Date: 18.05.2018

Meynard, J. M., Aggeri, F., Coulon, J. B., Habib, R., and Tillon, J. P. (2006). Recherches sur la conception de systèmes agricoles innovants. *Rapport du groupe de travail.*

Mueller, N. D., Gerber, J. S., Johnston, M., Ray, D. K., Ramankutty, N., and Foley, J. A. (2012) "Closing yield gaps through nutrient and water management", *Nature*, 490: 254. https://doi.org/10.1038/nature11420 Access Date: 15.11.2018

O'connell, P. F. (1992) "Sustainable agriculture-a valid alternative", *Outlook on Agriculture*, 21(1): 5-12.

Önen, Ö., Fındık Güvendi, G., and Beşeren, H. (2016) "Organoklorlu Pestisitlerin Yüksek Omurgalı Böbreği Üzerindeki Histopatolojik Etkileri - The Histopathological Effects of Organochlorine Pesticides on Kidney of Advanced Vertebrates", *Kafkas Üniversitesi Fen Bilimleri Enstitüsü Dergisi*, 9(2): 26-35.

Prather, M., Derwent, R., Ehhalt, D., Fraser, P., Sanhueza, E., and Zhou., X. (1994) Radiative forcing of climate change: Other trace gases and atmospheric chemistry. In E. Alyea, T. Bradshaw, J. Butler, M. . Carroll, D. Cunnold, E. Dlugokencky, ... D. Wuebbl (Eds.), Intergovernmental Panel on Climate Change (pp. 73-126). Cambridge University Press.

Reytar, K., Hanson, C., and Henninger, N. (2014) "Indicators of sustainable agriculture: A scoping analysis", Working Paper, Installment 6 of Creating a Sustainable Food Future. https://doi.org/10.1016/j.enpol.2011.05.043 Access Date: 05.07.2018

Rohila, A., Duhan, A., Maan, D., Kumar, A., and Kumar, K. (2017) "Impact of Agricultural Practices on Environment", *Asian Journal of Microbiology, Biotechnology and Environmental Sciences*, 19: 381-384.

Steinfeld, H., Gerber, P., Wassenaar, T., Castel, V., Rosales, M., and de Haan, C. (2006) Livestock's Long Shadow. FAO of the UN. Retrieved from http://www.fao.org/docrep/010/a0701e/a0701e00.HTM Access Date: 05.07.2018

Taşkaya, B. (2004) "Tarım ve Çevre", *TC Tarım ve Köy İşleri Bakanlığı Tarımsal Ekonomi Araştırma Enstitüsü TEAE-Bakış*, 5(1): 11-15.

Tsiafouli, M. A., Thébault, E., Sgardelis, S. P., de Ruiter, P. C., van der Putten, W. H., Birkhofer, K., and Hedlund, K. (2015) "Intensive agriculture reduces soil biodiversity across europe", *Global Change Biology*, 21(2): 973-985. https://doi.org/10.1111/gcb.12752 Access Date: 06.07.2018

Turhan, Ş. (2005) "Tarımda Sürdürülebilirlik Ve Organik Tarim", *Tarım Ekonomisi Dergisi*, 11(1): 13-24. http://dergipark.gov.tr/download/article-file/253316 Access Date: 06.07.2018

Weithmann, N., Möller, J. N., Löder, M. G. J., Piehl, S., Laforsch, C., and Freitag, R. (2018) "Organic fertilizer as a vehicle for the entry of microplastic into the environment", *Science Advances*, 4(April): 1–7. http://advances.sciencemag.org/content/advances/4/4/eaap8060.full.pdf Access Date: 06.07.2018

Yılmaz, A., Ünvar, S., Koca, T., and Koçer, A. (2017) "Türkiye'de Biyogaz Üretimi ve Biyogaz Üretimi İstatistik Bilgileri",*Technological Applied Sciences*, 12(4):218–232. http://dergipark.gov.tr/nwsatecapsci/issue/31523/339927 Access Date: 06.07.2018

Yurtseven, E., Çakmak, B., Kesmez, D., and Polat, E. (2010) "Tarımsal Atık Suların Sulamada Yeniden Kullanılması",*Türkiye Ziraat Mühendisliği VII. Teknik Kongresi*. http://www.zmo.org.tr/resimler/ekler/4e299e28c5847ef_ek.pdf Access Date: 18.05.2018

Metin Kılıç

An Outlook on Companies' Environmental Activities in terms of Corporate Governance

Introduction

With the industrial revolution, the success of capitalism created some opportunities for the enterprises to grow quickly. As a result of this growth, the investors brought their capitals together with the aim of providing enough resources to the big projects and established big companies. In line with their aims to obtain more benefits, these new companies used natural resources in an unlimited and unconscious manner for growth and profit without considering the harm that they gave to the environment.

The ecology movement on environmental pollution and energy saving that started in the 1960s spread with almost the same extent as the growth and globalization of the companies that paved the way for this movement. During the 2000s, personal and social consciousness on the environment became much more obvious. It can also be said that the environmental issues started to be used as a competition resource in the areas of business and politics. The companies that want to go a step further than its rivals have understood that not only profit, but also long-term social support is important in the presence of increasing environmental consciousness and movements.

Corporate governance has generally been accepted by all of the stakeholders as good and effective governance principles aiming to enable companies to function with a fair (or equal), transparent, responsible and accountable governance understanding. The significance of balancing the relations between company and stakeholders as well as those among stakeholders in a way that will contribute to corporate actions is emphasized, and the main principles of it are handled in order to achieve it.

This study analyzed the environmental consciousness and companies' actions within the four principles of "fairness (or equality), transparency, responsibility and accountability".

1 The Industrial Revolution

The Industrial Revolution is explained in four development processes as is seen Fig. 1.

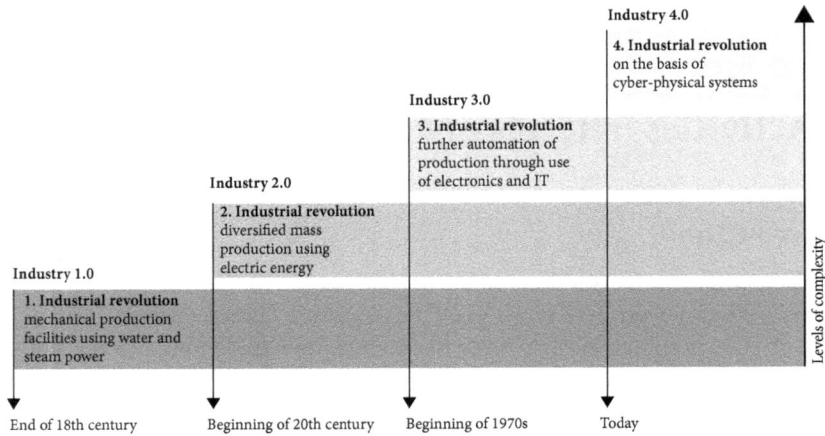

Fig. 1: Industrial Revolutions One to Four. Source: Buhr, 2017: 5.

The period that started with the invention of steam machine in the 18th century is defined as the first industrial revolution. Human and animal based production was replaced by machine based one. The first practices were made in textile industry in England (Küçükkalay, 1997: 52). The usage of coal and wood to support steam power in manufacturing, transformation to mass production, search for international market and raw material, urbanization and population increase can be mentioned as the most determining changes of this period (EBSO, 2015: 4).

The second industrial revolution can be said to have started with the invention of electricity. The usage of more developed machines in production came into play. The invention and usage of electricity replaced coal. In addition, mass production developed during this period. Transportation network, particularly the railway, enlarged. Ore usage (especially iron, steel etc.) in the industry increased and heavy industry was developed during this period (http://www.sanayidegelecek.com; EBSO, 2015: 5). With the usage of electricity in production and the development of transportation network, capital need for industrialization increased, and intercontinental communication led to the development of capital markets. During this period, the need for raw material usage and new markets increased, and petroleum and its variants became the primary raw materials.

The First and Second World Wars enabled companies and technologies to develop in the military area while causing social devastation and collapse of the empires. Economic Depression of 1929 devastated capital markets and world economies in a serious way. With the ending of the Second World War, the applications in the military area were transferred to the economic area, which paved the way for the third industrial revolution. As of the 1950s, digital technologies started to develop during the period called third industrial revolution (www.sanayidegelecek.com: EBSO, 2015: 4). The development of computer and communication technologies, the effects of nuclear energy and power, global polarization, development of capital markets and dependency on natural resources can be evaluated as the most important indicators of this period. The usage of information technologies especially in production processes created a process that also included personal product design, and competition increased in raw material and product market. This period can be expressed as a duration when global companies formed an empire.

During this period, the technological developments contributed to the formation of financial assets. The increase of market types related to financial markets and of the assets traded in these markets as well as the relevant transactions can be possible only with the emergence of new communication and computer systems. As the purchasing and selling parties of the financial markets aim to gain profit, financial innovations decrease the cost of purchase-sale transactions. Credit cards, international dimension of financial markets and securitization can be considered financial innovations emerging with the technological development. With the developments in computer technology, small investors can become a partner with big consistent investment tools and buy or sell securities at securities exchange (Kutukız, 2003: 121). It is possible to mention that this situation leads to the globalization of capital.

In the fourth period of the industrialization, the effect of heavy usage of information technologies is high. In particular, the fourth industrial revolution is characterized by: widespread and broadly accessible internet; smaller, cheaper and more powerful sensors; artificial intelligence and machine learning. The drivers of the change are physical (autonomous vehicles, 3d printing, advanced robots, new materials), digital (internet of things, relationship between things and people connected by technologies and platforms) and biological (genetic sequencing and genetic engineering, synthetic biology and biological editing) (Kuruczleki, Pelle, Laczi and Fekete, 2016: 327).

2 Environmental Effects of Companies and Environmental Movements

Negative effects of companies and the development periods of industries increase together. With the usage of wood and coal in production, demand for natural resources increased, which subsequently led to the unconscious consumption of other natural resources. Petroleum and other valuable minerals such as steel, aluminum, uranium, copper, gold and silver were continuously extracted and used as if they were limitless with the effect of technological developments, which became a significant factor for the inevitable emergence of environmental problems. This process caused pollution and degradation in land, water and air that are the most valuable things for human beings. For this reason, environmental movements emerged as a reaction to companies' such actions. As one of the important worldwide environmental disasters, we can mention wars, first atomic bomb, Bhopal, Chernobyl disaster, Seveso, Love Canal, Exxon Valdez, Aral Lake, Deepwater Horizon, (Lepisto, 2009; www.history.com; www.cnnturk.com; Deniz and Küçük, 2005: 1261; www.ntv.com; Dalkıranoğlu, 2016; Micklin, 1988: 1170; www.bbc.com).

The examples provided above refer to rare, but extreme and serious outcomes caused by companies. In fact, when companies' activities are examined, it is seen that they use chemicals, destroy tress and water resources while they are obtaining natural resources, that they use energy and create wastes while processing and storing them, that the shipping vehicles give harm to the environment during the distribution of products and pipe lines damage natural balance, that harms such as leakage and explosions occur in storehouses, all of which destroy the usage of environmental assets and affect the environment and hence the future generations negatively by shortening the life of sustainability, namely life. Apart from them, many other factors such as the damages occurring during the end-use process, air and water pollution resulting from coal burning, wastes remaining after product usage, piles of rubbish do the same effect. In light of these explanations, the damages that companies give to the environment are realized in the forms of air, water, land, noise pollution, wastes, erosion, ecocide (visual pollution, deforestation, pasture loss, wetland loss, shore pollution, decrease of biological diversity and habitus etc.).

The harms that companies give to the environment during their activities lead to irreversible outcomes. These harms cause depletion of ozone layer and climate changes, which makes the world to become an unlivable place. The social reaction against these negative outcomes on which companies have significant effects can be analyzed in two periods.

The period starting from the end of 19th century till the middle of 20th century during which the first social movements and organizations to protect the environment were active are called as the first wave environmental movements. Many nature organizations that still continue their existence in the 21st century were founded during that period. The determining feature of the first wave environmental movement is that it is directed toward the protection of natural living spaces, wildlife, wild animals, especially the birds, and old forests. The first wave was actually of European and North American origin. American nature protection is divided into two groups. In "preservationism" movement that aims to protect the nature by preventing human intervention, the nature has its own value and needs to be protected from all kinds of human intervention. The second movement of American protectionism refers to wise use, and the idea of using unthreateningly is dominant. According to this movement, sustainable management and wise use of natural resources is required for economic growth. Nature protection in Europe, on the other hand, developed in the forms of three waves as environmental outcomes created by the industrial revolution, environment consciousness thanks to the development in natural sciences and romantic movement developed against rationalist values of enlightenment idea. These effects were experienced differently at different parts of Europe. These movements enabled legal regulations for the protection of the nature. However, they stopped because of the First and Second World Wars. The most important achievements of the movements that revived after the wars are IUCN (the International Union for the Conservation of Nature), founded in 1949 with the aim of protecting the nature, and the international nature protection organization WWF (World Wildlife Fund), founded in 1961. The first wave movement aiming to protect the nature has selective characteristics and is pioneered by scientists, experts, technicians and wealthy people (Cerit Mazlum, 2014: 212–213).

The second wave ecology movement targets insatiable materialism of industrial society and over consumption. The ecology movement questions uncontrolled economic growth, centralized state and industrialization ideology that it considers the reasons of environmental devastation. In terms of this feature, the ecology movement differs from nature protectionism. It criticizes the industrial society both politically and culturally. The ecology movement demands open, decentralized, egalitarian, democratic and participatory social relations instead of bureaucratic, hierarchical, centralist, closed, sexist and unequal social and political order (Cerit Mazlum, 2014: 214).

In the birth and development of this movement, the effect of Rachel Cason's "silent spring", published in 1962, on the society with regard to the environmental and health problems was significant (www.lenntech.com).

The research that Roma Club, founded in 1968, got Massachusetts Technology Institute to carry out was published in 1972 with the title "Limits to Growth". This study foregrounded the significant and strong relationship between economic development and the environment and hence directed the attention to the environmental issues (Bozlağan, 2005: 1015). The book contains a message of hope, as well: Man can create a society in which he can live indefinitely on earth if he imposes limits on himself and his production of material goods to achieve a state of global equilibrium with population and production in carefully selected balance (www.clubofrome.org). UN Environment Conference, held in Stockholm during the same dates, was the first global step in this respect. After the Brundtland Report, in which the first official definition of sustainable development was made, Rio Conference was held in 1992, where strategies regarding the environment and development were analyzed in detail and the agenda of 21st century was determined. With Kyoto Protocol, the framework of the combat with global warming and climate change was identified. In UN Millennium Summit of 2000, Millennium Development Goals were determined under the auspices of UN. In 2002, on the other hand, the World Sustainable Development Summit was held in Johannesburg with the aim of finding out more effective sustainable development strategies to apply the decisions taken in Rio Conference (Aksu, 2011: 11).

This endeavor or the following ones show that there is an intersocietal understanding that a sustainable development will not be possible without having a sustainable environment. Especially the technological developments of the 21st century enabled people to give similar reactions. Knowledge can spread more quickly than before and individuals can take actions and give reactions regardless of social support. The significant effects of social pressures especially involve the positive changes in legal regulations. Many countries have identified some rules for the protection of environment so that it can be handed down to the future generations and a sustainable development is ensured, and serious sanctions have been imposed on those who do not comply with them. On the condition that laws are incapable, many environmental organizations put pressure on companies by activating individuals against companies and their products. In the face of legal obligations and social reactions, many companies attempt to increase their environmental conscience with a good or bad grace. Many companies carry out environmental actions voluntarily and inform the stakeholders with different methods that they desire especially with the aim of becoming a part of sustainable development and taking their support more by showing their environmental consciousness to the society.

3 Corporate Governance and Its Main Constituents

The notion of corporate governance entered into the literature in 1990. In the Report of the Committee on Financial Aspects of Corporate Governance, prepared under the auspices of Sir Adrian Cadbury and mentioned with his name, corporate governance is defined as "the system by which companies are directed and controlled". Cadbury Report explains the relations among the parts that form this system as well as handling with corporate governance as a system (www.ecgi.org).

"Boards of directors are responsible for the governance of their companies. The shareholders' role in governance is to appoint the directors and the auditors and to satisfy themselves that an appropriate governance structure is in place. The responsibilities of the board include setting the company's strategic aims, providing the leadership to put them into effect, supervising the management of the business and reporting to shareholders on their stewardship. The board's actions are subject to laws, regulations and the shareholders in general meeting" (Cadbury Report, 1992: 2–5)

The report also shaped many following practices, law and regulations. Shleifer and Vishny, on the other hand, define corporate governance as methods that ensure investors that provide fund to the company about the return that they will obtain from their investments (Shleifer and Vishny, 1996: 2). Other most significant studies on corporate governance refer to the corporate governance principles that were determined and announced by the OECD. The OECD corporate principles were determined in the OECD Council Meeting, held on 27–28 April 1998, upon the call of governments, interested international corporations and private sector representatives, for the development of a series of standards and leading principles related to corporate governance. The OECD Principles were first published in 1999 and updated in 2015 after some amendments by many organizations and individuals (OECD, 2015: 4, www.oecd.org/).

The corporate governance principles aim to help the OECD member and non-member governments to evaluate and develop legal, corporate and regulatory framework for corporate governance in their countries. In addition, these principles aim to guide other stock markets, investors, companies and corporations that have a role in the development of a good corporate governance process as well as making suggestions to them. These principles focus on the companies whose share certificates are traded at securities exchange. However, they can also be applied in the companies whose share certificates are not traded at securities exchange. For instance, they can be beneficial to develop corporate governance in private or public enterprises. These principles form a common ground that the OECD countries will take as a basis in developing good corporate governance applications. It is aimed to make these principles simple, understandable and accessible for the international community (OECD, 2004: 13, www.oecd.org/).

Corporate governance principles have been increasingly accepted as effective and good governance principles that enable long-term aims determined by considering the shareholders that affect or will affect company activities and by leaving the idea of determining the aims of company governance for short-term profit earning.

The attempt to deal with companies' insufficient accounting and auditing activities and their negative outcomes has been a significant factor in the development of corporate governance and its principles. Central position of companies in the society may cause different people and groups to behave in accordance with their own interests or expectations when it comes to company activities. When it is observed that social responsibility and environment conscience develop in parallel with corporate governance, it can be said that these reactions can start a process that will end with companies' closing down. The environmental effects of company activities on the society are important. For this reason, corporate governance principles take the attention of company directors with regard to considering environmental activities and their effects while carrying out their activities.

For an effective corporate governance application, the corporate governance application of a company is grounded on four main constituents. These main constituents are fairness (or equality), transparency, responsibility and accountability.

Fairness (or equality) implies fair (or equal) treatment of all shareholders and stakeholders in corporate governance and prevention of interest conflicts (SPK, 2005: 3). As required by this constituent, the directors of the company need to behave equally to all parts that are affected by the decisions and applications of the company while they are carrying out activities. This is a prerequisite of a fair and equal governance understanding (Demirbaş and Uyar, 2006: 24).

In companies, stakeholders do not only make contributions, but they also have some interests. In compliance with fairness, the corporate governance is expected to fulfill the interest of each stakeholder in accordance with the amount of his/her contribution (Kılıç, 2009: 132).

Transparency is the timely, correct, complete, clear, interpretable, cheap and easy accessible promulgating of financial and non-financial information related to the company, excluding trade secrets and the information that has not been announced to the public (SPK, 2005: 3). All the stakeholders want to know the company's financial situation, structure, the activities of directors and the board of directors as well as the interests of workers, customers, public and shareholders (Oliver, 2004: 6). Transparency emphasizes the need for fulfilling those wishes of the stakeholders.

Responsibility refers to the auditing of all activities that corporate management carries out on behalf of the company and their compliance with the regulatory agreement and intercorporate arrangements (SPK, 2005: 3). The duty of governing the company on behalf of shareholders has been transferred to the executives. The executives are directly responsible for all the activities of the company. The executives fulfilling these responsibilities need to comply with some legal arrangements such as current labor law, environment legislation and tax laws.

The constituent of responsibility emphasizes that managers are not only responsible to shareholders, but also all stakeholders, since the company does not only belong to shareholders. The corporate management is responsible to all stakeholders that affect or are affected by all stakeholders.

Accountability refers to the necessity of the members of the board of directors to be mainly accountable to joint stock company's corporate body and hence to shareholders (SPK, 2005: 3). Accountability is a constituent that implies that all persons who give decisions and take actions in the company are obliged to be liable and accountable for their decisions and actions (Kılıç, 2009: 135). Accountability emphasizes that the workers at each rank of the company are obliged to give an account of the outcomes of their actions in terms of their duties and liabilities to a higher rank. As far as company's executives are considered, it implies company's accountability to shareholders in general council while they are directing them toward the aims determined by the shareholders and to the state, relevant lawmakers in legal structure and the society.

The OECD's corporate governance principles resemble a building built on four main blocks. When the OECD's corporate governance principles are examined, it is seen that 19 principles or explanations make emphasis on the environment: They are explained below respectively:

1. "The Principles recognise the interests of employees and other stakeholders and their important role in contributing to the long-term success and performance of the company. Other factors relevant to a company's decisionmaking processes, such as environmental, anti-corruption or ethical concerns, are considered in the Principles… but are treated more explicitly in a number of other instruments including the OECD Guidelines for Multinational Enterprises, the Convention on Combating Bribery of Foreign Public Officials in International Business Transactions, the UN Guiding Principles on Business and Human Rights, and the ILO Declaration on Fundamental Principles and Rights at Work, which are referenced in the Principles" (OECD, 2015: 10).

In the OECD's corporate governance principles, it is emphasized that the environment is effective in enabling the long-term success of the company as well as the interests of all stakeholders and that the principles are determined accordingly.

2. The corporate form of organization of economic activity is a powerful force for growth. The regulatory and legal environment within which corporations operate is therefore of key importance to overall economic outcomes (OECD, 2015: 14)

The notion of environment here has been used in order to refer to the legal structure with which the enterprises are in relation. The main emphasis in the Principles is that a company cannot be successful without complying with the legal system. When it is evaluated in terms of the environment, it is emphasized that the compliance of the company with its country's environmental arrangements is significant for success.

3. Corporate governance requirements and practices are typically influenced by an array of legal domains, such as company law, securities regulation, accounting and auditing standards, insolvency law, contract law, labour law and tax law. Corporate governance practices of individual companies are also often influenced by human rights and environmental laws (OECD, 2015: 15).

Herein, it is directly mentioned that companies are influenced by the environmental laws. In line with this principle, it is emphasized that company activities need to be carried out by taking the environmental laws into consideration.

4. The rights of stakeholders are often established by law (e.g. labour, business, commercial, environmental and insolvency laws) or by contractual relations that companies must respect (OECD, 2015: 37).

This principle emphasizes stakeholders' rights and points out the necessity of the compliance with the environmental laws in ensuring the rights of stakeholders.

5. Disclosure also helps improve public understanding of the structure and activities of enterprises, corporate policies and performance with respect to environmental and ethical standards and companies' relationships with the communities in which they operate (OECD, 2015: 42).

In the disclosures of the company, it is required to explain the necessity of the compliance with environmental standards and hence of informing the stakeholders of the environmental issues.

6. In addition to their commercial objectives, companies are encouraged to disclose policies and performance relating to business ethics, the environment

and, where material to the company, social issues, human rights and other public policy commitments (OECD, 2015: 43).

    In this Principle, it is emphasized that identification and explanation of environmental policies in company activities is required.

7. Users of financial information and market participants need information on reasonably foreseeable material risks that may include: risks that are specific to the industry or the geographical areas in which the company operates; dependence on commodities; financial market risks including interest rate or currency risk; risk related to derivatives and off-balance sheet transactions; business conduct risks and risks related to the environment (OECD, 2015: 46).

    This principle classifies the risks that financial investors and companies may encounter and requires information on the probability of enterprisers to encounter these risks. It also accepts environmental risks as a risk element that may affect the decision of financial investors of the enterprise and mentions that the corporate management needs to make an explanation of the environmental risks that are effective risk elements in the decisions of investors.

8. Another important board responsibility is to oversee the risk management system and systems designed to ensure that the corporation obeys applicable laws, including tax, competition, labour, environmental, equal opportunity, health and safety laws (OECD, 2015: 51).

    This principle mentions the liabilities of the board of directors while emphasizing that the compliance with environmental laws is included in this liability. It is underlined that the board of directors is responsible for fulfilling the company activities in accordance with the environmental laws.

9. The board is not only accountable to the company and its shareholders but also has a duty to act in their best interests. In addition, boards are expected to take due regard of, and deal fairly with, other stakeholder interests including those of employees, creditors, customers, suppliers and local communities. Observance of environmental and social standards is relevant in this context (OECD, 2015: 51).

    In another principle, it is emphasized that the compliance of the board of directors with environmental standards, in line with the interests of all stakeholders, is under the responsibility of the board of directors.

10. "Other laws that may be applicable include those relating to taxation, human rights, the environment, fraud, and money laundering" (OECD, 2015: 56).

This principle implies that the board of directors is liable for complying with environmental laws.

As required by fairness, one of the constituents of corporate governance principles, the companies mustn't take actions that will affect the whole society negatively and must act in accordance with the expectations of stakeholders while considering the future of the company.

According to transparency, on the other hand, the information on the environmental risks, environmental policies of the executives as well as explanations on the compliance with environmental standards need to be provided to all stakeholders. The companies need to give information with the methods that they determine (in financial reports, independent environment reports, sustainability reports or integration reports).

According to responsibility, the company's board of directors is directly liable to determine the environmental risks in company activities, ensure compliance with environmental law and standards and provide explanation on these issues. The board of directors is both authorized and liable to carry out company's environmental activities in accordance with the laws and the expectations of stakeholders.

According to accountability, the board of directors is responsible for company's environmental activities. The board of directors can account for their duty to all stakeholders with a disclosure and report. In addition, with annual reports, environment reports, sustainability reports or integration reports, the companies notify and give account to the relevant parties. In annual general assemblies of the companies, the board can give account to the shareholders by presenting company activities and to the public with statutory audits.

The most attention-grabbing feature of corporate governance principles is that the nature of company's environmental activities isn't limited by some standards and principles. As a result, if the environmental laws are not of the required quality in the country where the company is operating, it may not be possible to expect attention and care from it. For the companies that act in accordance with corporate governance principles to comply with the corporate governance in terms of environmental issues, the country in which they are operating need to have applicable, auditable environmental law and standards that have sanctions.

Conclusion

The development of industry is based on natural resources. However, natural resources are scarce and limited. Companies are mainly enterprises that have been founded in order to provide profit to shareholders by producing, selling or mediating goods and services that will meet the needs of other companies or individual through natural resources. However, it has been understood that the irresponsible and limitless usage of natural resources especially during the third period of the industrial revolution resulted in more and much quicker ecocide. These reactions that were primarily realized as nature protection grew to be an ecologic movement toward the end of 20th century and became dominant worldwide. Many countries and corporations developed environmental laws or standards they would comply with in their company activities. Individuals have begun not purchasing the products and services of the companies that do not comply with these laws or standards, or the states punish these companies according to the laws.

The OECD corporate governance principles have been determined and suggested to be applied with the aim of ensuring good and efficient governance for companies, especially those that exist in capital markets. These principles are grounded on four main constituents. These are fairness or equality, transparency, responsibility and accountability. Corporate governance principles refer to companies' environmental activities. When these references are examined, company's display of sensitivity to environmental issues is a requirement of fairness or equality. In performing environmental activities and responsibilities of companies, the board is fundamentally responsible and has all the required authority to fulfill this responsibility. As a result, it is under the responsibility of the board to govern all processes in terms of company's environmental activities (determining the risks and environmental policies of the company and its compliance with environmental laws and standards. It is important for the company's board of directors to explain their method of fulfilling these responsibilities as well as the amount that they have fulfilled by means of disclosures or reports that they will prepare. Responsibility and transparency based understanding in corporate management requires the directors to be accountable. For this reason, accountability based governance will enable company directors to pay more attention to the laws and social reactions in their decisions and applications. The corporate governance principles foreground the compliance with the laws and standards. Therefore, among the priorities of the lawmakers in terms of companies' environmental activities, decreasing environmental problems must have a place. Otherwise, the corporate governance principles cannot by themselves prevent or eliminate environmental problems.

We should not forget that we do not have another world to live, and that it is both our and future generation's right to live in this world with the available resources under the best conditions as much as possible. Protecting this right is our most important duty.

References

Aksu, C. (2011) "Sürdürülebilir Kalkınma ve Çevre", Güney Ege Kalkınma Ajansı, Denizli,1–33.

Bozlağan, R. (2005) "Sürdürülebilir Gelişme Düşüncesinin Tarihsel Arka Planı", Sosyal Siyaset Konferansları Dergisi, 50: 1011–1028.

Buhr, D. (2017) "Social Innovation Policy for Industry 4.0, Bonn: Division for Social and files/wiso/11479.pdf, Access Date: 25.07.2018

Cadbury Report, (1992) "Report of the Committee on the Financial Aspects of Corporate Governance. Burgess Science Press", London, http://www.ecgi.org/codes/documents/cadbury: pdf, Access Date: 25.07.2018

Cerit Mazlum, S. (2014) "Çevrecilik ve Çevreci Kareketler", (Yeni Toplumsal Hareketler, Editör: Kartal, B. ve Kumbetoğlu, B.) Eskişehir: T.C. Anadolu Üniversitesi Yayın No: 2345,

Dalkıranoğlu, B. (2016) "İnsan Evladının Eliyle Yaptığı ve Çevresel Felaketlere Neden Olan 10 Korkunç Şey" https://listelist.com/insanlarin-neden-oldugu-felaketler, Access Date: 12.06.2018.

Demirbaş, M. and Uyar, S. (2006) "Kurumsal Yönetim İlkeleri ve Denetim Komitesi", İstanbul: Güncel Yayıncılık, 1. Baskı.

Deniz, V. and Küçük, S. (2005) "Afetler ve Endüstriyel Kazalar", Deprem Sempozyumu, 23–25: 1261–1263, Mart, Kocaeli.

EBSO (Ege Bölgesi Sanayi Odası), (2015) "Sanayi 4.0",http://www.ebso.org.tr/ebsomedia/documents/sanayi-40_88510761.pdf, Access Date: 25.07.2018.

Kartal, B. and Kumbetoğlu, B. (2014) "Yeni Toplumsal Hareketler", Eskişehir: Anadolu Üniversitesi Yayını, Yayın No: 2345.

Kılıç, M. (2009) "Kurumsal Yönetim", Senem Besler (Editör) Yönetim Yaklaşımlarıyla Kurumsal Sürdürülebilirlik, İstanbul: Beta Yayınları, 1. Basım.

Kuruczleki, E., Pelle, A., Laczi, R. and Fekete, B. (2016) "The Readiness of the European Union to Embrace the Fourth Industrial Revolution," Management, University of Primorska, Faculty of Management Koper, 11 (4): 327–347.

Kutukız, D. (2003) "Finansal Yeniliklerin Gelişimi, Piyasalar Üzerindeki Etkileri ve Türkiye Üzerinde Deneysel Bir Çalışma", Mali Çözümler Dergisi, 65: 114–126. https://www.ismmmo.org.tr/Yayinlar/Mali-Cozum-Dergisi/sayi-65/--2082, Access Date: 25.07.2018

Küçükkalay, M. (1997) "Endüstri Devrimi ve Ekonomik Sonuçlarının Analizi", Süleyman Demirel Üniversitesi İktisadi ve İdari Bilimler Fakültesi Dergisi, 2: 51–68.

Lepisto, C. (2009) "8 Worst Man Made Enviromental Disasters of All Time", https://www. treehugger.com/natural-sciences/8-worst-man-made-environmental-disasters-of-all-ti me.htm l, Access Date: 10.06.2018.

Micklin, P. P. (1988) "Desiccation of the Aral Sea: A Water Management Disaster in the Soviet Union", Science, 241(4870): 1170–1176.

OECD Principles of Corporate Governance, (2004) http://www.oecd.org/corporate/ca/corpor ategovernanceprinciples/31557724.pdf, Access Date: 25.06.2018.

OECD Principles of Corporate Governance (2015) https://www.oecd.org/daf/ca/Corporate-Governance-Principles-ENG.pdf, 1–61, Access Date: 25.06.2018.

Oliver, R.W. (2004) "What is Transparency?", New York: McGraw-Hill; 1. Edition,

SPK, (Sermaye Piyasası Kurulu) (2005) "Kurumsal Yönetim İlkeleri", http://www.spk.gov.tr/Sayfa/Dosya/66, Access Date: 26.06.2018.

Shleifer, A. and Vishny R. W., (1996) "A Survey of Corporate Governance", National Bureau of Economic Research, Working Paper No: W5554.

https://www.history.com/topics/atomic-bomb-history, Access Date: 12.06.2018.

https://www.cnnturk.com/turkiye/cernobil-faciasi-neydi-turkiyeyi-nasil-etkilemisti?page=1, Access Date: 12.06.2018.

https://www.ntv.com.tr/galeri/yasam/en-buyuk-10-cevre-felaketi,qRjTxCun4Uuv2NgA_THzGQ/Hus LVI4kvkW8N44u7gP-2w, Access Date: 12.06.2018.

https://www.bbc.com/turkce/haberler/2010/09/100908_bp_leak.shtml, Access Date:12.06. 2018.

http://www.sanayidegelecek.com/sanayi-4-0/tarihsel-gelisim/, Access Date: 07.06.2018.

https://www.lenntech.com/development-environmental-movement.html, Access Date: 20.06. 2018.

https://www.clubofrome.org/report/the-limits-to-growth/, Access Date: 20.06.2018.

Mustafa Gül

Environmental Accounting

Introduction

The world faces many environmental problems due to industrialization, rapid population growth and unplanned urbanization. Environmental problems such as air, water and noise pollution, global warming, the growing depletion of the ozone layer, reduction of green zones, glacier melting, seasonal changes and acid rains threaten the health and future of the world. As people have become aware of this situation, they have started to request businesses to offer goods and services with the most benefits in addition to respecting and cherishing the environment in which they live. This also caused countries to form international solidarity and cooperation. In this context, everyone has important duties in the light of the "think globally, act locally" philosophy. The objectives of the conventional business administration were to achieve economic growth and make profits while current businesses aim to achieve sustainability and keep the environment clean and healthy within the framework of environmentally conscious management mentality as part of the corporate social responsibility.

Businesses must carry out their activities without damaging the environment, eliminate damages they cause on the environment, if any, minimize the damages if they are not able to eliminate them completely and do whatever necessary to make up for the damages they cause. Businesses will have to incur some costs in connection with such activities. Costs that are incurred by businesses in connection with environmental activities are called environmental costs in the literature. Such costs are either ignored or completely included in general production costs in conventional accounting and cost accounting. This prevents businesses incurring environmental costs from determining the environmental costs. It is highly important to know what each environmental cost is and for which product it is incurred in terms of cost accounting and management accounting that help to make managerial decisions. Furthermore, the concept of full disclosure requires that environmental costs must be conveyed to information users by means of reports so that they can obtain information on this matter. Thanks to environmental accounting that takes environmental costs into account, businesses can make more accurate and healthier administrative decisions and information users can be provided with accurate information about activities of businesses.

1 Business – Environment Relationship within the Scope of Social Responsibilities of Businesses

Like all other entities, businesses are economic units that are both affected by and affect the environment in which they live in various ways. Therefore, current and future policies of businesses have to comply with the requirements of the environment and be created in line with such requirements (Karalar, 1995: 41). That is because the environment is the common property of all entities in the world. So business managers have to take not only their own interests but also the environment in which they live into account while making decisions. Otherwise, they cannot maintain their activities. That is because the external environment of the business will not allow it.

Social responsibility represents accountability of businesses in terms of business activities affecting people, the society and environment. This view suggests that factors affecting people and the environment negatively must be certainly improved (Öz-Alp, 1996: 46). The concept of social responsibility includes relationships of businesses with their external environments. The main areas of responsibility of businesses can be listed as follows (Dinçer, 1998: 158–162):

- Social values and business ethics,
- Not abusing their power,
- Protecting the natural environment,
- Protecting consumers,
- Improving the quality of the work life,
- Making investments that increase social benefits,
- Other social aids and supports.

Businesses reduce the quantity and quality of soil, air, water and other natural resources by using them during their activities in various ways as well as contaminating the air, water and soil with their waste. As part of their social responsibilities, businesses must protect the environment in which they live, take measures in order not to damage the environment and carry out all necessary activities to compensate any damage they cased, if any.

In fact, social responsibility is an attempt of social agreement and reconciliation. Social agreement is based on efforts for safe products for public welfare and happiness, realistic and limited amount of commercials, safe work places for employees, activities that do not disrupt the environment or jeopardize the natural life, equal treatment of everyone and creating employment opportunities. Economic duties and responsibilities of businesses cannot be considered separately from their social duties and responsibilities (Eren, 2004: 48). They

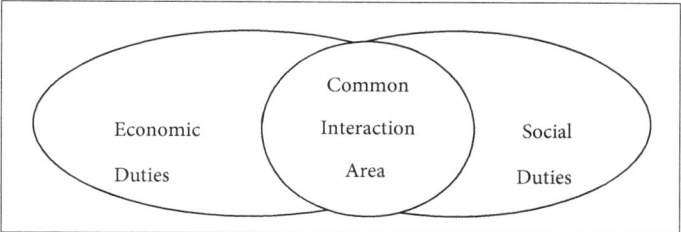

Fig. 1: The Common Interaction Area of Businesses. **Source:** Eren, 2004: 49.

are intertwined as can be seen Fig. 1; they must be organized together, and their mutual interaction must be taken into consideration. In other words, the duty of businesses is not to deal with merely economic activities but to fulfill their abovementioned social responsibilities toward the environment in addition to economic activities.

Environmental awareness that has spread all over the world has also affected Turkey significantly. Businesses turn to environmental protection activities especially due to concerns about globalization. The fact that the European Union attaches utmost importance to environment forces businesses to carry out such activities (Sabuncuoğlu and Tokol, 1997: 30–31).

Environmentally conscious management is a mentality adopted by businesses that take the natural environment into account as an important element during decision making processes, aim to minimize or completely eliminate damages to the environment during their activities, attempt to place the philosophy of protecting the ecological environment that changes product design and packaging as well as production processes in the corporate culture in this context and fulfill their duties toward the society within the scope of social responsibility (Nemli, 2000–2001).

As you can see in the Tab. 1, the aim of the conventional management mentality is to achieve economic growth and make profits whereas the aim of environmentally conscious management mentality is to achieve sustainable development and life quality.

2 Environmental Accounting Concept

Various definitions have been made for environmental accounting that is also known as green accounting in the literature. Environmental accounting is an

Tab. 1: Comparison of the Conventional Management Mentality and Environmentally Conscious Management Mentality. Source: Nemli, 2000–2001

	Conventional Management	**Environmentally Conscious Management**
Aims	• Economic growth and profit • Revenue for partners	• Sustainability and life quality • Welfare of partners
Products	• Products designed for function, style and price • Packaging that creates redundant waste	• Environmentally friendly products designed for the environment
Organization	• Hierarchical structure • Top-down decision making • Centralization in decision making	• Non-hierarchical structure • Participative decision making • Decentralization in decision making
Environment	• Dominating the environment • Managing the environment as a resource • Regarding pollution and waste as externalities	• Being in harmony with the nature • Realizing that natural resources are not unlimited • Management and minimization of pollution and waste
Operating functions	• Aims to increase marketing and consumption. • The financier wants to maximize profits in the short term. • Accounting department focuses on conventional costs. • Human resources management aims to increase employee productivity.	• Marketing exists for consumer education. • The financier aims for long term sustainable growth. • Accounting department focuses on environmental costs. • Human resources management aims to achieve occupational health and safety.

umbrella concept with three different areas of use: national income accounting, financial accounting and management accounting (EPA, 1995: 4) As you can see in Tab. 2, environmental accounting at country level is considered within the context of national income while environmental accounting at business level is considered within the context of financial and management accounting.

2.1 Environmental Accounting in the Context of National Income Accounting

Environmental Accounting within the scope of National Income Accounting is an accounting system at the macroeconomic level. Gross Domestic Product

Tab. 2: The Scope of the Environmental Accounting Concept. Source: EPA, 1995: 4

Types of Environmental Accounting	Objective	Target Group
1) National Income Accounting	Country	Outside the business
2) Financial Accounting	Business	Outside the business
3) Management / Cost Accounting	Business, department, facility, product line or system	Inside the business

(GDP) is an example of this. GDP mostly indicates economic conditions of people. The environmental accounting term is used in this sense as well. For instance, environmental accounting can be used in monetary and physical units. In this sense, environmental accounting is also called natural resource accounting (EPA, 1995: 4). Statistics about the value, quality and amount of both renewable and non-renewable natural resources and the consumption of the country are called natural resources accounting (EPA, 1995: 28). The pioneer of this approach is the Department of Social Resources established by the Soviet Union in 1974. This department created and developed the Natural Resources Accounting. Following these developments, the French government developed a similar accounting system in 1978 and it was followed by the Canadian government (Pearce ve Markandya, 1993: 89).

2.2 Environmental Accounting at Business Level

Various definitions have been made for environmental accounting, which is also known as green accounting in the literature. Some of these definitions are as follows:

A broad-based term that refers to the incorporation of environmental costs and information into a variety of accounting practices (Snapshots of Environmental Cost Accounting A Report To: US Epa Environmental Accounting Project, 1998: 8).

Environmental accounting is the act of accounting financial incidents related to environment and showing them in financial statements (Aslan, 1995: 22).

Environmental accounting refers to inclusion of all environmental costs in financial reports of businesses (http://www.ea.gov.au/industry/eecp/tools/tools4.html).

Defined as a subcategory of accounting, environmental accounting is a series of activities including explanation, analysis and recording of economic impacts of an environmental activity or environmental problems of a defined economic system (Schaltegger and Burritt, 2000: 63 reported in Bilen and Seyitoğulları, 2016: 1744).

Environmental accounting aims to achieve sustainable development, maintain a favorable relationship with the community and pursue effective and efficient environmental conservation activities. These accounting procedures allow a company to identify the cost of environmental conservation during the normal course of business, identify benefits gained from such activities, provide the best possible means of quantitative measurement (in monetary value or physical units) and support the communication of its results (Environmental Accounting Guidelines, 2002: 3).

If we try to make a definition similar to these definitions in terms of bookkeeping, we can define it as "recording, classifying, summarizing and reporting environmental transactions that are of financial quality in monetary units and interpreting the results (Kırlıoğlu and Can, 1998): 56).

2.2.1 Environmental Accounting in the Context of Financial Accounting

Financial accounting defines, measures and creates information showing the financial status and activity results of a business and transforms it into information that can be used by relevant parties by analyzing and interpreting it (Cemalcılar and Önce, 1999: 1). Financial reports that will be submitted to relevant people or institutions and prepared in accordance with generally accepted accounting concepts and principles must include environmental liabilities and financially material environmental costs (EPA, 1995: 4). Accordingly, costs incurred in connection with all kinds of environmental activities must be registered and included in the accounting information system. This way, financial accounting provides environmental cost accounting with the information it needs. Although they are different from each other in terms of their functions and tasks, there is a constant exchange of information between them.

2.2.2 Environmental Accounting in the Context of Management Accounting

Management accounting provides useful accounting information for business management. The aim of this accounting branch is to collect and analyze accounting data that will help business managers plan and check business actions and make decisions about special cases (Üstün 1997: 8). Environmental costs and performances of businesses must be taken into account in calculation and analyses of costs that will guide the senior management through their decisions (EPA, 1995: 5).

Management accounting uses a broad set of cost and performance data by business managers in making many business decisions while management

Tab. 3: Obtaining Various Managerial Decision Benefits from Environmental Costs. Source: EPA 1995: 6.

✓ Product design	✓ Capital investments
✓ Process design	✓ Cost control
✓ Facility deployment	✓ Waste management
✓ Purchase	✓ Cost allocation
✓ Activity	✓ Product mix
✓ Risk management	✓ Product pricing
✓ Environmental compliance strategies	✓ Performance evaluation

accounting in the context of environmental accounting refers to the use of data about environmental costs and performance in business decisions and operations (EPA, 1995: 5). Using environmental cost information while making various managerial decisions such as product design, capital investments, process design, cost control, facility deployment, waste management, purchase, cost allocation, activity and product mix, risk management, product pricing, environmental compliance strategies, performance evaluation etc., as shown in Tab. 3, will be useful for the business.

Information needed by environmental accounting in the context of management accounting can be obtained by developing environmental cost accounting. So firstly the environmental costs and what they mean for businesses must be determined.

3 Environmental Costs

Environmental costs can be defined as all sacrifices made by businesses in connection with their environmental activities. Environmental activities refer to transactions and applications carried out by businesses in order to avoid damaging the environment and/or make up for damages that have been caused.

It is important to define and explain environmental costs related to products, production processes, systems or facilities so that the business management can make the right decisions. Achieving objectives such as reducing environmental costs, increasing income and improving environmental performance requires the necessary importance to be attached to current and possible future environmental costs. Definition of an environmental cost by a business depends on how it will use this information (cost allocation, capital budgeting, process/product design and other management decisions etc.), scope and level of the activity (EPA, 1995: 7).

There are a few ways of understanding whether a cost is an environmental cost. There is no doubt that costs incurred in order to obey environmental laws are certainly environmental costs. Costs related to improvement of the environment and costs of pollution control equipment are certainly accepted as environmental costs. Environmental protection and improvement activities that are determined legally as well as activities that are not mentioned in the law must be regarded as environmental costs (EPA, 1995: 11).

There are other costs but they are in the gray area. In other words, we do not know for sure if they are environmental costs. For example, is the use of clean technologies or efficient energy turbines an environmental cost? Are materials that become waste after checking shelf lives of raw materials environmental costs? It can be hard to distinguish some environmental costs from health and safety costs or risk management costs (EPA, 1995: 12).

Some companies use the following approaches to distinguish costs in the gray area (EPA, 1995: 12):

- Accepting a cost as environmental cost for only one purpose,
- Accepting the cost of an activity as partly environmental or
- Accepting a cost as environmental for accounting purposes when the business decides that more than 50 % of this cost is environmental.

Environmental costs are also considered to be one of the costs incurred by businesses to produce goods and services. On the other hand, environmental performance has become one of the criteria showing success of businesses in protecting the natural environment. Environmental costs and environmental performance are deemed important for business management for the reasons listed below (Özbirecikli and Melek, 2002: 83).

- Most of the environmental costs can be reduced to a great extent or completely eliminated as a result of decisions made by the business management regarding matters such as investment in a greener production process, redesigning the investment/production process and/or the product etc. rather than short-term changes in their activities.
- Such costs can be monitored in the general expenses account or even ignored due to important reductions that can be made in environmental costs.
- Many companies can make up for environmental costs by acquiring clean technology licenses or selling by-products that are considered to be waste.
- Better management of environmental costs has an important place in providing significant benefits for human health and increasing environmental performance in addition to the success of a business.

- Knowing environmental costs and environmental performance related to the production process helps to determine product costs and prices more accurately and design products, services and production processes of companies in a way that is more environmentally preferable.
- Products, services and production processes that are environmentally preferable for customers provide competitive advantages.
- Accounting of environmental costs and environmental performance can support improvement of a company and activities of the whole environmental management system (EMS). Such a system has become a necessity for companies with international connections for the near future (due to ISO 14001).

From the viewpoint of a business, such benefits cannot be ignored. Therefore, it is important for businesses to attach due importance to environmental activities so that they can survive, improve their profits and, most importantly, protect the natural environment.

3.1 Classification of Environmental Costs

The environmental cost concept has two basic dimensions: Societal Costs and Private Costs (EPA, 1995: 1).

3.1.1 Societal Costs

Societal costs are costs for which the business is not held accountable to individuals, the society or the environment (EPA, 1995: 1).

Various methods are used to determine societal costs. These are (Özbirecikli 2002: 51):

- Reduction (Avoidance) Cost Method,
- Loss Cost Methods,
- Usage Cost Methods.

3.1.2 Private Costs

Private costs are costs that affect profits or losses of a business directly (EPA, 1995: 1). In other words, costs incurred by businesses to avoid damaging the environment, minimize and compensate any damages are defined as private costs. So all kinds of prices paid as money by businesses for the environment are accepted as private costs.

Terms such as complete, total, real and life cycle etc. are used in the environmental accounting terminology to emphasize that environmental costs (and/or income) have been included in the accounting system. That is because the field of activity is incomplete as conventional approaches ignore private environmental costs (and cost saving and earnings) (EPA, 1995: 7)

Environmental cost items that are usually ignored or completely included in the general production cost account must be determined so that healthier managerial decisions can be made. This part of the study makes a general definition and classification of private costs that are included in the environmental cost accounting. This classification is given in Tab. 4.

Hidden Costs: Environmental costs that are overlooked or accumulated in general production pools are called hidden costs (EPA, 1995: 36).

Regulatory Costs: They include costs incurred due to legal regulations made by local or national authorities in connection with environmental regulations (EPA, 1995: 35).

Voluntary Costs: They are costs which are not required or necessary for compliance with environmental laws but voluntarily incurred by a business. Such costs result from precautions and activities beyond legal regulations (EPA, 1995: 35).

Upfront Costs: These costs include pre-production costs incurred for processes, products, systems and facilities (EPA, 1995: 35).

Back-End Costs: They include environmental costs that arise following the useful life of processes, products, systems or facilities. They are also called exit costs (EPA, 1995: 35).

Contingent Costs: They are environmental costs that are not certain to occur in the future but depend on uncertain future events. Sometimes they are referred to as environmental liabilities, liability costs or contingent liabilities (EPA, 1995: 36).

Image and Relationship Costs: They are costs incurred by businesses to develop and maintain relationships with the environment, local community, customers, suppliers, investors, lenders and the public with the aim of creating and improving the company image. They are also referred to as less tangible costs (EPA, 1995: 37).

When some environmental costs are recorded in general production costs within the scope of environmental accounting and the products for which they are recorded are not known, such costs must be allocated to relevant production costs accurately. For example, when environmental costs are imposed on a product although they are not related to the product or environmental costs incurred for a product are imposed on this product incompletely, production costs cannot be calculated accurately.

Tab. 4: Classification of Private Costs. Source: EPA, 1995: 9

Potentially Hidden Costs		
Regulatory Costs	**Upfront Costs**	**Voluntary Costs (Beyond Compliance)**
▸ Notification	▸ Site studies	▸ Community relations
▸ Reporting	▸ Preparation costs	▸ Monitoring/testing
▸ Monitoring/testing	▸ Permitting	▸ Training
▸ Studies/modeling	▸ R&D costs	▸ Audits
▸ Remediation	▸ Engineering costs	▸ Qualifying suppliers
▸ Recordkeeping	▸ Procurement costs	▸ Reports (e.g., annual environmental reports)
▸ Plans	**Conventional Costs**	▸ Insurance
▸ Training	▸ Capital equipment	▸ Planning
▸ Inspections	▸ Materials	▸ Feasibility studies
▸ Manifesting	▸ Labor	▸ Remediation
▸ Labeling	▸ Supplies	▸ Recycling
▸ Preparedness	▸ Utilities	▸ Environmental studies
▸ Protective equipment	▸ Structures	▸ R&D
▸ Medical surveillance	▸ Salvage value	▸ Habitat and wetland protection
▸ Environmental insurance	**Back-End Costs**	▸ Other environmental projects
▸ Financial assurance	▸ Disposal of inventory	▸ Financial support to environmental groups and/or researchers
▸ Pollution control	▸ Post-closure care	
▸ Spill response	▸ Site survey costs	
▸ Waste management		
▸ Taxes/fees		
	Contingent Costs	
▸ Future compliance costs	▸ Remediation	▸ Legal expenses
▸ Penalties/fines	▸ Property damage	▸ Natural resource damages
▸ Response to future releases	▸ Personal injury damage	▸ Economic loss damages
	Image and Relationship Costs	
▸ Corporate image	▸ Relationship with professional staff	▸ Relationship with lenders
▸ Relationship with customers	▸ Relationship with workers	▸ Relationship with host communities
▸ Relationships with investors	▸ Relationship with suppliers	▸ Relationship with regulators
▸ Relationship with insurers		

Conclusion

Survival in the current global market has started to be directly proportional to the importance attached to the environment. This forces businesses to adopt an environmentally conscious management mentality. Naturally, it is inevitable for environmental protection activities carried out within the scope of this management mentality to impose additional costs on businesses. It is highly important that such costs, which are also referred to as environmental costs, are determined, recorded, classified, analyzed and interpreted so that effective managerial decisions can be made. Environmental accounting is required to achieve this. In order to record and report environmental costs within the scope of environmental accounting, it is highly important that businesses create a separate account group or account class such as an "environmental costs" account showing each environmental cost as well as the location of such costs. They must also prepare and implement an account plan. Furthermore, we think that it is important to create relevant accounting standards aimed at preventing different calculations of environmental costs, which are indispensable in this context, at each business and to create and implement an account plan in accordance with these standards so that environmental cost reports of businesses can be analyzed, compared and monitored by the state.

References

Aslan, Ü. (1995) "Çevre Muhasebesi ve Nuh Çimento AŞ.'inde Çevre Muhasebesi Üzerine Pilot Bir Çalışma", Yayınlanmamış Yüksek Lisans Tezi, Anadolu Üniversitesi Sosyal Bilimler Enstitüsü, Eskişehir.

Bilen A. ve Seyitoğulları O. (2016) "İş Örgütlerinde Çevre Muhasebesi Algısına Yönelik Bir Araştırma: Diyarbakır İli Örneği", İnsan ve Toplum Bilimleri Araştırmaları Dergisi, 5: 7.

Cemalcılar, Ö. ve Önce, S. (1999) "Muhasebenin Kuramsal Yapısı", Eskişehir: Anadolu Üniversitesi Yayınları No: 1093.

Dinçer, Ö. (1998) "Stratejik Yönetim ve İşletme Politikası", Beşinci Basım. Beta.

Environmental Accounting Guidelines (2002), https://www.env.go.jp/en/policy/ssee/eag02.pdf Access Date: 20.07.2018.

EPA (Environmental Protection Agency), (June 1995) "An Introduction to Environmental Accounting as A Business Management Tool: Key Concepts and Terms", Washington, D.C.: United States Environmental Protection Agency, https://www.epa.gov/sites/production/files/2014-01/documents/busmgt.pdf Access Date: 25.07.2018.

Eren, E. (2004) (Editör Necdet Timur), "Stratejik Yönetim" 1. Baskı, Eskişehir: AÖF Yayın No: 801.

Hughes, S.B. and Willis, D.M. (1995) "How Quality Control Concepts can Reduce Environmental Expenditures", Journal of Cost Management, 9(2): 15–19.

Karalar, R. (1995) "İşletme: Temel Bilgiler İşlevler", Dördüncü baskı. Eskişehir: ETAM A.Ş.

Kırlıoğlu, H. ve Can, A.V. (1998) "Çevre Muhasebesi", 1. Basım Adapazarı: Değişim Yayınkları.

Nemli, E. (Ekim 2000–Mart 2001) "Çevreye Duyarlı Yönetim Anlayışı" İ.Ü. Siyasal Bilgiler Fakültesi Dergisi, 23–24: 211–224.

Öz-Alp, Ş. (1996) "İşletmelerin Sosyal Sorumlulukları," Anadolu Üniversitesi Öğretim Fakültesi Dergisi, 2(1):41–50.

Özbirecikli, M. (2002) "Çevre Muhasebesi" Birinci basım, Ankara: Natürel Kitap ve Yayıncılık.

Özbirecikli, M. ve Melek, Z. (2002) "Çevre Muhasebesi ve Çevresel Maliyetlerin Maliyet Muhasebesine Etkileri ve Bir Araştırma," Muhasebe ve Finansman Dergisi, 14: 82–91

Pearce, D. ve Markandya, A. (1993) "Yeşil Ekonomi İçin Mavi Kitap" Çeviren: Türksen Kafaoğlu ve Arslan Başer Kafaoğlu İstanbul: Alan Yayıncılık, Birinci Baskı.

Sabuncuoğlu, Z. ve Tokol, T. (1997)" İşletme I-II", Bursa:Uludağ Üniversitesi.

Sevilengül, O. (1997) "Tekdüzen Muhasebe Sistemi İle Uyumlu Genel Muhasebe", 6. baskı, Ankara: Gazi Kitapevi.

Schaltegger, S. and Burritt, R. (2000) Contemporary Environmental Accounting: Issues Concepts and Practice. Greenleaf Publishing, Sheffield.

Snapshots of Environmental Cost Accounting A Report To: US EPA Environmental Accounting Project, (May 1998) http://www.dep.state.pa.us/dep/deputate/pollprev/pdf/ALLSNAPS.pdf Access Date: 14.07.2004.

Üstün, R. (1997) "Yönetim Muhasebesi" Üçüncü baskı, İstanbul: Bilim Teknik Yayınevi.

Meltem Ece Çokmutlu and Metin Kılıç

Evolution of Environmental Reporting: The Example of Turkey

Introduction

Rapid population growth, unplanned urbanization and industrialization, unlimited use of science and technology to enhance life standards of humanity, natural disasters, rapid globalization etc. cause a series of environmental problems such as air, water, soil, noise and visual pollution etc. (Haftacı and Soylu, 2008: 93, Çalış, 2013: 175). Environmental awareness and environmental protection have increased since the 1960s in order to reduce and prevent environmental problems. All national and international institutions and organizations have started to act in an environmentally conscious manner and take environmental factors into account during decision making processes. Social pressures, consumer demands, various environmental organizations and legal regulations have also caused businesses to take action about the environment in the national and international arenas (Haftacı and Soylu, 2008: 93, Türk and Erciş, 2017: 353). Today, societies fiercely object to the fact that businesses carry out their activities by imposing social, environmental and economic costs on the society in order to maximize their own interests (Aktekin, 2014: 6).

Businesses take precautions in order to minimize damages they have caused or might cause on the environment in line with their responsibilities (Çalış, 2013: 175). Awareness raising activities have revealed that precautions taken against environmental damages must be shown in economic values. Considered to be a free good, the environment has started to be regarded as a commodity and businesses affected by environmental developments have started to take environmental factors into account during decision making processes. One of the objectives of businesses is to make profits and they have acknowledged that they have to attach importance to the environment in their production methods and investment plans and started to carry out activities aimed at protecting the environment as they have realized that they might not be able to find an environment where they can do business and make profits in the medium and long term if they fail to attach due importance to the environment (Haftacı and Soylu, 2008: 93).

Businesses in Turkey just like many countries and many businesses operating in these countries have started to take responsibilities for protecting the environment and especially public businesses have started to prepare reports to inform stakeholders about their sensitivity to the environment. These efforts of businesses have brought "environmental accounting and environmental reporting" into question (Kaya and Varıcı, 2008: 209).

1 The Necessity for Accounting Environmental Costs and Environmental Accounting

The environment is important for businesses to exist and maintain their existence. Environmental factors affect businesses by limiting them and creating opportunities and competitions for them. Businesses which obtain inputs necessary to continue their activities and realize their objectives from the nature can survive as long as they provide the outputs requested and accepted by the market. Otherwise, businesses will face problems such as legal sanctions, social sanctions, resource shortage or cost increase. As a result, goods and services offered as outputs must possess qualities that will be accepted by the market (Güven, 2013: 27–29). It can be said that customers in many markets have been making decisions regarding their product preferences by taking impacts of businesses and products on the environment into account in recent years. This causes business managers to pay more attention to environmental activities and businesses to realize environmental costs.

International Accounting Standards Committee (IASC) defines environmental costs as the costs of efforts necessary to manage environmental impacts of a business. The committee proposes that environmental factors must be recorded by businesses (Ergin and Okutmuş, 2007: 148).

Milne regards environmental costs as costs incurred in order to physically measure environmental impacts of human activities and protect the environment (Milne, 1991: 91), Graff and Reiskin define the same concept as costs incurred by a business to eliminate problems related to its production process affecting the environmental quality (Graff and Reisikin, 1998).

Accounting of environmental costs within other cost categories will prevent businesses from monitoring environmental activities and preparing environmental reports. However, if we consider environmental costs separately from other costs, decisions related to matters such as product mix, selection of product costs, evaluation of pollution prevention projects, waste utilization, comparison of environmental costs, product pricing, production design, stopping production etc. can be made more clearly (Murray, 1989: 404).

If businesses can determine environmental costs and reflect them in reports accurately, they can both fulfill some legal liabilities and achieve other potential benefits from environmental accounting practices. Providing the production process, system and product with added value can be given as an example of this. This way, businesses will have a privileged position and obtain competitive advantages as goods and services will be preferred more by consumers/customers (Çalış, 2013: 182).

Businesses calculate environmental costs and report them to users in order to minimize pressures from interest groups such as environmental protection organizations, investors and customers and acquire an "environmentalist business" image in the society, thus strengthening the prestige of the business and enabling the management to manage environmental risks better (Çalış, 2013: 181).

In short, businesses have to take environmental consequences of their activities in addition to economic consequences into account due to the increase in social awareness and roles of environmental organizations and environmental legislations. Resources to be used by businesses in order to achieve this goal, the resulting costs, transactions related to accounting and reporting of such costs reveal relationships with accounting (Güven, 2013: 39).

2 Reporting Methods of Environmental Activities

Corporate reporting refers to notification of the financial and non-financial status of a business to information users on a regular basis in print and/or electronically (Uyar, 2015: 1). The scope of corporate reporting went through a considerably big change and improvement toward the end of the 20th century. Businesses used to collect only financial information through financial statements, but they have been reporting non-financial information in the last 30 years (Uyar, 2015).

Different reporting methods have been adopted due to lack of a standard regulation on this matter despite the interest of businesses in environmental reporting. One of these methods is that businesses extend the scope of their financial reports and annual activity reports and explain environmental information within the scope of these reports while the other method refers to independent environmental reports prepared separately from financial reports (Ulusan, 2009: 185, Martin and Hadley 2008: 245, Reported in Kaya and Varıcı, 2008: 209, Güven, 2013: 80). Another method is sustainability reporting showing financial, environmental and social responsibilities of a business together and the latest method is integrated reporting that takes into account the impacts of institutions on the future, and possible impacts of financial, environmental and social activities on the future value of the business. Accordingly, environmental activities of businesses can be presented to interest groups through four types of corporate reports. These reports are financial reports, independent environmental reports, sustainability reports and integrated reports.

Businesses can prepare such reports in accordance with their priorities and qualities of the corporate structure. However, the environmental reporting process indicates an order as mentioned above.

2.1 Financial Reporting

We can say that financial reporting is the first step of corporate reporting (Saban et al., 2017: 916). Today many international businesses explain environmental information in their activity reports (Ulusan, 2009: 185). The environmental report is submitted only in the content part, end note part or both content and end note part of the financial report when the annual report extension method is used.

The scope of financial reporting has expanded toward social aspects since the 1970s. The main reason for the scope extension of financial reporting is the intention of submitting information to internal and external stakeholders about company activities, products, services and relevant positive and negative social impacts. This first reporting attempt which is aimed at presenting non-financial information focuses on social aspects or social impact and social efficiency to some extent. Environmental reporting emerged around 1990 and extended the scope of the initial social reporting activities significantly (Herzig and Schalttegger, 2006: 304–305).

2.2 Independent Environmental Reporting

Independent environmental reports (IER) are prepared by business administrators who argue that it is not possible to include sufficient information about the environment in annual reports and even if it is possible, specific attention cannot be paid to the environment (Taşdemir, 2011: 92–93).

IER can be defined as system outputs used to transfer environmental information to information users. Businesses prepare and publish IER as a means of showing their environmental commitment clearly and providing information about their environmental performance and attempts (Ulusan, 2009). IER is a part of the information system that transfers environmental information about a business to interest groups (Yağlı, 2006: 84). These reports prepared by businesses will have significant benefits in analysis of environmental costs. The success of such reports is directly related to the support of the senior management, participation of the accounting department and understandability of the reports. These reports are obtained from values integrated with the accounting system and they can be prepared on a daily, weekly, monthly or annual basis in line with the requirements (Ergin and Okumuş, 2007: 156).

Businesses usually make explanations about the Environmental Policy, Environmentally Friendly Products, Environmental Rewards, Environmental Management System Certificate, Environmental Costs, Environmental Training Given to Employees, Environmental Auditing, Waste Control and Protection of Environmental Damage (Uyar, 2015: 54).

Many environmental reporting guides have been developed to encourage businesses to collect environmental information. Environmental reporting guides are designed in a way that businesses that want to report their environmental performance to the public on a regular basis can use them as a reference while preparing environmental reports. Guides address a series of problems related to environmental reporting including structures and contents of environmental reports. These guides are usually in the form of checklists for contents of environmental reports including both quantitative and qualitative information. Many environmental reporting guides have been published since the early 1990s and they are used by businesses in various countries (Ulusan, 2009: 186–191).

2.3 Sustainability Reporting

The fact that conventional financial reporting and IER do not take social impacts of business activities into account has been their main shortcoming in terms of sustainability and businesses have started to submit more detailed information related to economic, social, environmental, managerial and financial performance through sustainability reports and integrated reports (Tüm, 2014: 67, Aras and Sarıoğlu, 2015).

Corporate sustainability reporting that emerged in the 2000s as corporate social responsibility reporting and has improved up until now (sustainability reporting in short) is a reporting method aimed at submitting information to report users about economic, environmental and social activities of businesses as well as two-dimensional connections between them (Önce, Onay and Yeşilçelebi, 2015: 235).

As existing corporate reporting tools including notification about environmental activities have complicated structures that are mostly based on financial capital, retroactive, short term, do not have integrated information and contain limited explanations, businesses prefer sustainability reports and integrated reports, which are currently considered to be the next stage of sustainability reports (IIRC, 2011; Yüksel, 2017: 125).

2.4 Integrated Reporting

Integrated reporting differs from other corporate reports as it submits activity results transparently to all stakeholders by taking all capital elements into account, using information in financial reports and sustainability reports and carrying out activities of the business in the short, medium and long term through strategies connecting the past and the future, and it is considered to be the ultimate level of reporting (IIRC, 2011: 9, Yüksel, 2017: 124, Saban, Vargün and Gürkan, 2017).

Integrated reporting is a brief explanation of how the strategy, corporate management, performance and expectations of a business can create values in the short, medium and long term in the external environment of the business. Integrated report is a corporate reporting tool that enables transfer of the value resulting from business activities based on integrated thinking to interest groups (Yüksel, 2017: 28). One of the capital elements explained in the International Integrated Reporting Framework is natural capital. Natural capital is the capital element including renewable and nonrenewable natural resources such as air, water, soil, metals and ecosystem used by a business in goods and services (Yüksel, 2017: 72–73). Interest groups are notified about environmental activities in detail and as integrated with other capital elements through integrated reporting which is considered to be the ultimate level of corporate reporting.

2.5 Comparison of Methods Used for Reporting Environmental Activities

Tab. 1 shows a comparison of similarities and differences of annual activity reports, environmental reports, sustainability reports and integrated reports that are used to make notification about environmental activities.

3 The Development Process and Current Status of Environmental Reporting in Turkey

When we look at the development process of environmental policies in Turkey, we see that Turkey, which was going through new period of change after World War II, entered a period of planned development in order to eliminate the imbalance between supply and demand. Development and industrialization efforts were

Tab. 1: Comparison of Annual Activity Report, Independent Environmental Report, Sustainability Report and Integrated Report. Source: Uyar, 2015: 5, Yüksel, 2017: 128.

	Annual Activity Report	Independent Environmental Report	Sustainability Report	Integrated Report
Content	Reported without establishing any connection between the financial status and business activities.	Qualitative and quantitative data related to environmental activities of the business	Information about the environmental, social and economic performance of the business is reported.	Information about the financial, environmental, social and economic performance of the business is reported in a way that a connection is established between the information.
Basic Capital Element	Financial and Intellectual Capital	Natural Capital	Financial capital, social and relational capital, human capital, intellectual capital and productive capital	Social and relational capital, natural capital, human capital and intellectual capital
Period	Short Term (1 year)	Short Term (1 year)	Short Term (1 year)	Short, Medium and Long Term
Obligation Status in Turkey	Compulsory	Voluntary	Voluntary	Voluntary
Regulatory Organization and Regulations in Turkey	Legislation	Environmental Reporting Guides	GRI - Sustainability Reporting Guides	IIRC - International Integrated Reporting Framework
Transparency and Accountability	Transparency at the legislation level	Limited transparency	Limited transparency	High transparency

(continued on next page)

Tab. 1: (continued)

	Annual Activity Report	Independent Environmental Report	Sustainability Report	Integrated Report
Matter in Focus	Financial performance and activity results of the previous period	Environmental activity results of the previous period	Environmental, economic and social performance results of the previous period	Strategies and results related to financial and non-financial capital elements establishing a connection between the previous period and the next period
Assurance	Legislation, internal auditing and independent external auditing	Internal auditing and optional independent external auditing	Evaluation and internal auditing in line with GRI reporting principles, ISO 14001 Environmental Standard and ISO 26000 Social Responsibility Standard, UN Global Compact, Carbon Disclosure Project	Evaluation and internal auditing in line with legislation, GRI reporting principles, ISO 14001 Environmental Standard and ISO 26000 Social Responsibility Standard, UN Global Compact, Carbon Disclosure Project

included in a plan with the Constitution adopted in 1961 and the development plans prepared by the State Planning Organization were put into practice as of 1962 (Sencar, 2007: 108). This rapid change in Turkey which entered a period of planned economic development had the most negative impacts on the environment. The first two plans prepared by this organization for the period between 1963 and 1972 did not have parts specific to environmental problems and no detailed policies were prepared about the environment. Only the "environmental health" issue was mentioned in general (Sencar, 2007: 110). Environmental problems were first addressed in the environmental problems section of the third five-year development plan that was prepared in 1973. Environmental problems were attached special attention in the plan in parallel with the environmental awareness resulting from the effects of the Stockholm Conference held in 1972. The first organizational structure established with the aim of preventing environmental pollution was the Prime Ministry Undersecretariat of Environment. Environmental awareness increased after adoption of the industrial ecology approach in the mid-1980s and total quality approach in the early 1990s. Preventive policies aimed at protecting the environment started to be prepared with the ongoing development plans and progress reports the first of which was prepared in 1998. Laws, regulations and by-laws related to environment were adopted in the following years. After the establishment of the Prime Ministry Undersecretariat of Environment, which was the first organizational structure regarding the environment in 1978, the name of the Ministry of Environment that was founded in 1991 was changed as the Ministry of Environment and Forestry in 2003, and it ultimately became the Ministry of Environment and Urban Planning in 2011. Today, organizations such as TUBITAK, Chambers of Industry and Commerce, Environmental Problems Foundation of Turkey, TEMA etc. in addition to public authorities carry out activities regarding environmental problems (Sencar, 2007: 110–133, Türk and Erciş, 2017: 354, Şenocak, 2017: 21–22).

Developed countries have made significant progress in environmental accounting and environmental reporting. Environmental reporting is voluntary in Turkey and there is no legal obligation. Therefore, it has not reached the standard of developed countries yet and it is not used as prevalently as in developed countries (Ulusan, 2009).

Studies on environmental accounting, environmental reporting and sustainability reporting in the Turkish literature indicate that information provided by businesses has improved both quantitatively and qualitatively (Ulusan, 2009). Especially corporations have been publishing their environmental awareness activities in their reports and informing interest groups on this matter lately. Such reports on environmental activities were referred to as environmental reporting in the 1990s whereas their scope was extended to include sportive, educational and cultural activities and they were called social responsibility reporting. Today such notification is commonly made within the scope of institutional sustainability or sustainability reporting (Uyar, 2015: 52). A review of the development process of notifications about environmental activities offered to interest groups has revealed the following:

ISO 14001 Environmental Management Standards are important standards that guide countries through fulfillment of their environmental responsibilities. They act as an important tool for ensuring sustainability and proving reliability of not only countries but also businesses and organizations (Şenocak, 2017: 28–29). A study conducted on 28 businesses that had acquired the ISO 14001 certificate in Turkey by 2002 revealed that environmental issues were not prioritized, and the mentality of environmental accounting and reporting was still new (Kaya, 2002).

Activities related to determination and recording of environmental costs and relevant reporting started in the early 2000s in Turkey (Kaya, 2002, Ergin and Okutmuş, 2007, Haftacı and Soylu 2008, Yıldıztekin, 2009, Ulusan, 2009, Çalış, 2013).

According to the regulation made by the Capital Markets Board in 2005, exchange companies were obligated to prepare a "Corporate Governance Compliance Report" in their activity reports or websites and submit it to the authorities. Article 17 in section three of this report titled "Stakeholders" stipulates that businesses need to make a statement about their activities regarding environment, the area in which they are located and the public in general as well as lawsuits filed against the business due to damages caused on the environment and the results of these lawsuits. On the other hand, despite this regulation that has been in effect since 2005, the main problems of environmental reports prepared in Turkey are as follows: the number of environmental reports prepared by Turkish businesses is low, reports concentrate on general and qualitative information and only good news are reported (Kaya and Varıcı, 2008: 225).

A review of environmental activities carried out in the early 2000s by businesses in Turkey reveals that businesses are not sensitive to the environment. It is observed that these businesses that are not sensitive to the environment do not pay sufficient attention to environmental accounting and environmental reporting. It has been found out that the main reason why businesses avoid environmental accounting and environmental reporting is that their accounting infrastructures are not suitable (Taşdemir, 2011: 161).

One hundred businesses out of the first five hundred businesses determined by Istanbul Chamber of Commerce were randomly selected and annual activity reports and independent environmental reports of these businesses were analyzed. It was determined that environmental reporting rate was lower than developed countries, businesses mostly used independent environmental reports and the number of businesses with ISO 14001 certificate was low (Kaya and Varıcı: 2008).

Environmental protection is considered to be a cost element in Turkey. However, expenses made on the environment in recent years indicate that realizing economic development by taking the environment into account and protecting the environment that is regarded as a cost element increase the competitive power and ensure sustainability (Sencar, 2007).

The number of businesses that prefer disclosing the economic, environmental and social dimensions of their activities to the public through institutional sustainability reports has increased in recent years. A total of 72 institutions in Turkey published 181 reports between 2005 and 2014 (01.05.2015). 130 of these reports were based on the GRI guide. As it can be understood from these numbers, the number of published sustainability reports has increased in years and the sustainability mentality is taken into account by more institutions each passing day (Önce, Onay and Yeşilçelebi, 2015).

It can be said that awareness about reporting and preparing non-financial information started to develop in 2007. The main focuses of sustainability practices in Turkey are shaped in line with business areas open for improvement. The main focuses of businesses that have changed from time to time since businesses started to focus on sustainability are determined in accordance with public policies and priorities of countries. Environment, reduction of energy costs, energy efficiency, gender equality and education are the main focuses of businesses today. Renewable energy consumption, poverty reduction based on regional development and disability areas are weak focus points (Gücenme Gençoğlu and Aytaç, TISK, 2016).

A study analyzing environmental and social applications based on activity reports of businesses registered at the sustainability index of Istanbul Stock Exchange found out that information about environmental sustainability included in annual activity reports increased until 2015. This shows that activities of businesses included in the environmental sustainability analysis and the importance attached to the environment have increased (Gücenme Gençoğlu and Aytaç, 2016).

It has been suggested in studies that it can be useful if regulatory institutions create a framework in line with sustainability objectives and a standard is developed for measurement and reporting of sustainability performance in Turkey (Şentürk and Fındık, 2015).

Five integrated reports were published by 2018 in Turkey (Arguden Governance Academy, Industrial Development Bank of Turkey, Aslan Çimento, Adana Çimento and Çimsa). It is understood from these reports that businesses accept the environment as a capital element (Önder, 2018). This can be an indicator of the fact that some businesses in Turkey have done more than just accounting and reporting environmental activities and the value attached to the environment has increased although it is a slow increase, and they have started to do the right things.

Conclusion

Other than natural causes, human activities are the real reason for environmental problems. Therefore, solution of environmental problems mostly depends on structuring human activities harmful to the environment in a way that they do not damage the environment. This is also a requirement of the environmental integrity principle of sustainable development.

Businesses in Turkey just like many countries and many businesses operating in these countries have started to take responsibilities for protecting the environment and especially public businesses have started to prepare reports to inform stakeholders about their sensitivity to the environment. These efforts of businesses have brought "environmental accounting and environmental reporting" into question. The number of businesses that report their environmental and social performances is rapidly increasing in Turkey just like in the world. These businesses, which aim to comply with public disclosure and transparency principles, use various reporting styles such as financial reports, annual activity reports, corporate social responsibility reports, sustainability reports and integrated reports while making notification about their environmental activities. When we look at the reports published by businesses in Turkey, we can say that social reports and social responsibility reports are used in a way similar to sustainability reports and contents of reports are not different from each other. However, the excessive amount of reports and the fact that information is presented in different reports without forming a connection have caused information users to distance themselves from the holistic point of view. Financial and non-financial information related to businesses has started to be presented in an integrated manner by forming connection between them in order to eliminate this problem.

The ultimate form of corporate reporting is known as integrated reporting. Although integrated reporting is still in its infancy, businesses see the environment as a capital element in line with integrated reporting and they share environmental statements with information users quantitatively and qualitatively.

Awareness must be raised in businesses by professional institutions and organizations related to environmental accounting in order to introduce environmental accounting and environmental reporting to businesses and make these businesses use them so that feasibility of corporate reporting can be increased in Turkey and healthier results can be achieved at businesses that use them. Standards must be determined by the state for environmental accounting and reporting, businesses preparing reports must be encouraged and environmental accounting must be taught at universities to train experts.

References

Aktekin, S. (2014) "Entegre Raporlama Türkiye'de Uygulanabilirliğinin Değerlendirilmesi ve Uygulama Örnekleri", Sermaye Piyasası Kurulu, Ortaklıklar Finansmanı Dairesi, Yeterlik Etüdü, Ankara.

Aras, G. and Sarıoğlu, G. (2015) "Kurumsal Raporlamada Yeni Dönem: Entegre Raporlama", TÜSİAD Yayınları, No: T/2015, 10:567.

Çalış, Y. E. (2013) "Çevresel Maliyetlerin Muhasebeleştirilmesi" Marmara Üniversitesi, İİBF Dergisi, 34(1): 175–190.

Ergin, H. and E. Okutmuş. (2007) "Çevre Muhasebesi: Çevre Maliyetleri ve Çevre Raporlanması", Yönetim Bilimleri Dergisi 5(1): 144–169.

Gücenme Gençoğlu, Ü. and Aytaç, A. (2016) "Kurumsal Sürdürülebilirlik Açısından Entegre Raporlamanın Önemi ve BIST Uygulamaları", Muhasebe ve Finansman Dergisi, 72: 51–66.

Güven, Z. (2013) "Çevre Maliyetleri Muhasebe Sistemi ve Bir Uygulama", Okan Üniversitesi, İşletme Anabilim Dalı, Yayımlanmamış Yüksek Lisans Tezi, İstanbul.

Haftacı, V. and K. Soylu. (2008) "Çevresel Bilgilerin Muhasebesi ve Raporlanması", Kocaeli Üniversitesi SBE Dergisi, 15: 92–113.

Herzig, C. and Schaltegger, S. (2006) "Reporting External Accounting Frameworks and Benchmarking: Corporate Sustainability Reporting", An Overwiev, Sustainability Accounting and Reporting (Ed: S. Schaltegger, M. Bennet and R. Burrit), Springer, Netherlands, Dordrecht, 301–324.

IIRC, (2011) Towards Integrated Reporting: Communicating in the 21st Century. http://theiirc.org/wp-content/uploads/2011/09/IR-Discussion-Paper-2011_spreads.pdf Access Date: 10.01.2019

Kaya, U. (2002) "İşletme-Doğal Çevre İlişkilerinin Mali Tablolar Aracılığıyla Raporlanması ve Denetimi", Karadeniz Teknik Üniversitesi, Sosyal Bilimler Enstitüsü, Yayınlanmış Doktora Tezi, Trabzon.

Kaya, U. and Varıcı, İ. (2008) "Gelişmekte Olan Ülkelerde Çevresel Raporlama: Türkiye Örneği", MÖDAV, 4: 209–227.

Martin, A. D., and Hadley, D. J. (2008) "Corporate environmental non-reporting–a UK FTSE 350 perspective", Business Strategy and the Environment, 17(4), 245–259.

Milne, M. J. (1991) "Accounting Environmental Resource Valves and Non-Market Valuation Techniques for Environmental Resources", A Review Accounting, Auditing Accountability Journal, 80–108.

Murray, D. and Frazier, K.B. (1989) "A Within-Subjects Test of Expectancy Theory in a Public Accounting Environment", 24(2):400–404.

Önce, S. Onay, A. and Yeşilçelebi, G. (2015) "Kurumsal Sürdürülebilirlik Raporlaması ve Türkiye'deki Durum", Journal of Economics, Finance and Accounting, 2(2): 230–252.

Önder, Ş. (2018) "Kurumsal Raporlamanın Yeni Trendi Entegre Raporlama", Ekin Yayınevi, Bursa.

Saban, M. Vargün, H. and Gürkan, S. (2017) "Yatırımcılara Bilgi Sağlama Aracı Olarak Entegre Raporlama", Muhasebe Bilim Dünyası Dergisi, 19(4): 915–936.

Sencar, P. (2007) "Türkiye'de Çevre Koruma ve Ekonomik Büyüme İlişkisi. Trakya Üniversitesi, Sosyal Bilimler Enstitüsü", Yayımlanmamış Yüksek Lisans Tezi, Edirne.

Şenocak, B. (2017) "İşletmelerde Çevresel Sürdürülebilirlik Bilinci, Denizli Tekstil İşletmelerine Yönelik Bir Araştırma", Pamukkale Üniversitesi, Sosyal Bilimler Enstitüsü, İşletme Anabilim Dalı, Yayımlanmamış Yüksek Lisans Tezi, Denizli.

Şentürk, F. and Fındık, H. (2015) "Türkiye'deki Akademik Dergilerde Çevre Muhasebesi Alanında 2006–2014 Yılları Arasında Yayınlanmış Bilimsel Makalelerin İçerik Analizi", Journal of Accounting, Finance and Auditing Studies, 1(3): 173–204.

Taşdemir, V. (2011) "İşletme Çevre İlişkilerinin Muhasebe Açısından Raporlanması", Ankara Üniversitesi, SBE, İşletme Anabilim Dalı, Yayımlanmış Yüksek Lisans Tezi, Ankara.

Tüm, K. (2014) "Kurumsal Sürdürülebilirlik ve Muhasebeye Yansımaları: Sürdürülebilirlik Muhasebesi", İnönü Üniversitesi Akademik Yaklaşımlar Dergisi, 5(1): 58–81.

Türk, B. and Erciş, A. (2017) "Türkiye'de Çevre Politikası ve Uluslararası Çevre Sözleşmeleri", International Journal of Social Science, 54: 351–362.

Türkiye İşveren Sendikaları Konfederasyonu (TİSK) ve Proje Ortakları (2016) "Herkes İçin Kurumsal Sosyal Sorumluluk Projesi", Türkiye Sürdürülebilirlik Raporlaması Ulusal İnceleme Raporu.

Ulusan, H. (2009) "Çevresel Raporlama Rehberleri ve İşletme Çevresel Raporlarında Açıklanması Gereken Bilgiler", Süleyman Demirel Üniversitesi, İİBF Dergisi, 14(2): 181–206.

Uyar, A. (2015) "Kurumsal Raporlamanın Gelişimi ve Güncel Yaklaşımlar", Gazi Kitabevi, Ankara.

Yıldıztekin, İ. (2009) "Sürdürülebilir Kalkınmada Çevre Muhasebesinin Etkileri", Atatürk Üniversitesi Sosyal Bilimler Enstitüsü Dergisi, 13(1): 367–390.

Yağlı, F. (2006) "Çevre Muhasebesi ve Mermer İşletmeleri Uygulaması: Ermaş Madencilik Tur. San. ve Tic. A.Ş. Örnek Uygulaması", Üniversitesi Sosyal Bilimler Enstitüsü. Yayınlanmamış Yüksek Lisans Tezi, Muğla.

Yüksel, F. (2017), "Entegre Raporlama", Ekin Yayınevi, Bursa.

Demet Beton Kalmaz

Economic Growth and Environmental Degradation in Turkey

Introduction

Global warming started to attract attention both in the academic literature and the policy arena in recent years, since there have been an increase in the risk of global warming arising as a result of industrial revolution transforming the global economy from organic economies based on human and animal power to inorganic economies based on fossil fuels. The main contributor to global warming is accepted to be the increase in Greenhouse Gases (GHG) through trapping heat and generating GHG effect in Earth's atmosphere. The GHGs in atmosphere include carbon dioxide (CO_2), methane (CH_4), nitrous oxide (C_2O), chlorofluorocarbons (CFC), hydrofluorocarbons (HCFC) and sulphur hexafluoride (SF_6). The main properties of GHGs are summarized in Tab. 1 below. Information about the anthropogenic sources, the concentration, global warming potential (GWP) and the time that the gas remains in the atmosphere per each GHG are provided.

CO_2 is regarded as the main contributor to global warming having the highest heat-trapping potential among other GHGs in the atmosphere (Özmen, 2009; Aytun & Akın, 2015), with the highest concentration ratio of 280,000 per billion. Global Warming Potential (GWP) is taken as constant in CO_2 remaining 100 years in the atmosphere (Center for Climate and Energy Solutions, 2017). The main causes of CO_2 occurrence are fossil-fuel combustion, land use conversion and cement arising from production. Increased production is called economic growth. Economic growth is highly related to energy consumption since energy use is essential for production. The energy demands of the countries have been increasing mainly because of the accelerated industrialization, technological improvement and urbanization to meet the development goals to achieve continuous production for sustainable economic growth. However, overuse or misallocation of energy resources generates high levels of CO_2 emissions.

Turkey is a developing country with US$863.712 billion gross domestic product (GDP) and US$10,891 per capita GDP according to the World Bank records of 2016. The share of exports of goods and services in GDP is 21.96 and

Tab. 1: GHGs and their Main Properties. Source: Center for Climate and Energy Solutions, 2017

	Anthropogenic Sources	Concentration (parts per billion)	GWP	Lifetime in Atmosphere (year)
Carbon Dioxide (CO_2)	Fossil-fuel combustion, Land-use conversion, Cement production	280,000	1	100
Methane (CH4)	Fossil fuel, Rice paddies, Waste dumps	700	25	12
Nitrous Oxide (C2O)	Fertilizer, Combustion, Industrial processes	270	298	114
Chlorofluoro Carbons (CFC)	Liquid coolants, Foams	0.534	10.900	100
Hydrofluoro Carbons (HCFC)	Refrigerants	0.218	1.810	12
Sulphur Hexafluoride (SF6)	Dielectric fluid	0.00712	22.800	3.200

imports is 24.85 %. As an emerging economy, Turkey has a current account deficits above 3.7 % of GDP. The main source of the current account deficit problem is considered to be the heavy dependency on imported energy resources over the years. Energy needs which leads to an increase in both energy consumption and production have been increasing constantly in all sectors in Turkey as a result of the increasing rate of population and urbanization (Bilgen et al., 2008). More than 75 % of the energy needs in total energy consumption is met from foreign energy sources imported from abroad (World Bank, 2018), which is an indication of the level of imported energy dependency of Turkey. The share of total energy sources in Turkey for 2015 is illustrated in Fig. 1 below.

Hydro, biofuels/waste and geothermal/solar/wind are renewable energy sources while natural gas, oil and coal are non-renewable sources. The highest share of the energy sources used in Turkey belongs to natural gas and coal, whereas the lowest belongs to renewable energy sources even if they are counted all together. Furthermore, coal and natural gas have the highest negative impact on environmental degradation since they have the highest ability

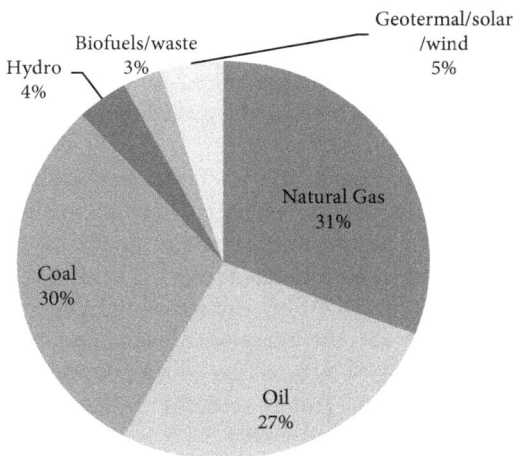

Fig. 1: Share of Energy Sources in Turkey in 2015. Source: Center for Climate and Energy Solutions, 2017.

to produce GHG, thus CO_2 emissions which stimulate environmental degradation (Keskingöz & Karamelikli, 2015) and global warming having harmful effects on ozone layer (Papadimitriou, 2004), which creates drawbacks on human health such as visual impairment, skin cancer and immune system impairment through letting harmful sun rays to reach earth (Onat et al., 2004). The highest share of natural gas and coal do not only decreases the environmental quality but also worsens the current account deficit in Turkey, since those sources are imported mostly from abroad as mentioned before. Out of the total energy consumption, more than 75 % were imported from abroad in 2015 (World Bank, 2018), while 99 % of natural gas was imported from Russia, Iran, Algeria and Nigeria, and more than 90 % of oil supplies was imported mostly from Iraq, Iran and Russia; accounting for the shares of total oil supplies in Turkey of 41 %, 20 % and 11 %, respectively (Republic of Turkey Energy Market Regulatory, 2017).

Most of the developed countries increase the use of more environmental friendly energy sources by applying energy policies to divert production by using more renewable energy sources instead of non-renewable. However, Turkey's non-renewable energy use – which is mostly import based sources – has been showing an increasing trend ever since; even though she is gifted with renewable energy sources of wind, solar, biomass, geothermal and hydro energy. Being a developing country with a very strong foreign energy and therefore import dependency leading to a persistent current account deficit accompanied with increasing levels of environmental degradation, Turkey is one of the most interesting countries to be focused both in the academic literature and the policy arena of the related topic.

This chapter of the book aims to give brief information about economic and environmental conditions and related policy applications in Turkey. To do so, the rest of this study follows with literature review summarizing the literature and the common and different results obtained from the investigation of the impact of Gross Domestic Product (GDP) on environmental degradation in Turkey, while the next subtitle presents the intergovernmental and binding agreements to reduce degradation, followed by a section which provides information about environmental degradation in Turkey and the policies and agreements to improve environmental quality and finally the final section of the study is the conclusion.

1 Literature Review

As the causes and consequences of global warming started to gain high attention, there have been several studies to investigate the impact of economic growth on environmental degradation. Econometric investigation of the debate is carried either by applying time series data for country case studies or by employing panel analyses for a group of countries. The results obtained from the studies in academic literature vary according to the data covering different time span or frequency, econometric methods employed and due to the fact that different case studies concentrating on different countries with varying characteristics of quantities and qualities of energy sources, political institutions, cultures and existing energy policies etc.

The impact of economic growth on environmental degradation in literature is concentrated in three different dimensions. The first tests the validity of the Environmental Kuznets Curve (EKC) theory to investigate the impact of economic growth on environmental degradation (Grossman & Krueger, 1995; Coondoo & Dinda, 2006; Lee & Lee, 2009). EKC hypothesis states that there is a U-shaped relationship between economic growth and CO_2 emissions, which indicates that a rise in per capita income increases CO_2 emissions during the early stages of economic growth and after a threshold is reached the increase in economic growth decreases CO_2 emissions. There are studies especially focusing on the validity of EKC for the case of Turkey (Katircioglu, 2017; Akbostanci, et al., 2009).

The second dimension aims to explain the relationship between economic growth and energy consumption, since the dominant factor in pollution is accepted to be the energy consumption through burning fossil fuel. Most of those studies followed Kraft and Kraft in 1978 to investigate the relationship by testing for co-integration and causality among the indicators (Narayan & Smyth, 2008; Yuan et al., 2007; Çetin and Ecevit, 2015; Hossain, 2011; Hossain, 2012; Chang, 2010; Cowan et al., 2014; Akin, 2014; Mohapatra & Giri, 2015; Salahuddin et al., 2015, Kesgingöz & Karamelikli, 2015; Altıntaş, 2013; Aytun & Akın, 2015; Katircioglu, Katircioglu and Altinay, 2017). The last dimension of the studies applies most recent techniques combining the first two by introducing energy consumption and some other economic indicators also into the EKC hypothesis to overcome the omitted variable bias of the model (Kasman & Duman, 2015; Halıcıoğlu, 2009; Çetin & Ecevit, 2015; Katircioglu, 2017; Koçak, 2014; Yavuz, 2014; Lean & Smyth, 2010; Acaravcı & Öztürk, 2010, Kivyiro & Arminen, 2014; Tang & Tan, 2015; Jalil & Mahmud, 2009; Kesgingöz & Karamelikli, 2015; Katırcıoğlu & Katırcıoğlu, 2018; Katırcıoğlu & Taspinar 2017; Pata, 2018; Bozkurt & Okumuş, 2015; Öztürk & Acaravcı, 2010 Katırcıoğlu & Çelebi, 2017; Özataç et al., 2017).

Test results vary among the studies in literature. Some of those are summarized by Tab. 2 below. Mostly the studies concentrating on Turkey are included to be able to compare the varying results since this chapter of the book aims to give information about Turkey. In addition, studies selected for concentrating on different countries to be able to provide worldwide information about the issue under consideration. The abbreviations used in the table to describe indicators are given in full names below the table.

2 Intergovernmental and Binding Agreements to Control Environmental Degradation

Climate change is a global problem seriously affecting human life. Therefore, several world-wide organizations and countries have been attempting to control environmental degradation through reducing GHG, especially CO_2 emissions, legally and most importantly starting with the agreement of the United Nations Framework Convention on Climate Change (UNFCCC).

UNFCCC is an international environmental treaty adopted in 1992 and entered into force when sufficient number of countries signed in 1994. Main objective of the treaty is "to achieve stabilization of GHG concentrations in the atmosphere at a level that would prevent dangerous anthropogenic interference with the climate system". The given time agreed to achieve such a level is decided

Tab. 2: Summary from the Literature. Source: Created by Author

Authors	Country	Indicators	Model	Co-integration	Result
Acaravcı & Öztürk (2010)	19 European Countries	EG, EC	Bounds Coint., Granger Causality Test	Yes (only for the case of countries)	Differing Results
Akbostancı et al. (2009)	Turkey	EG	Johansen Coint.	Yes	EKC is not valid
Akin (2014)	85 countries	EG, EC, TO	Pedroni/Kao/Fisher/FMOLS/ DOLS Coint, Granger Causality Test	Yes	$CO_2 \to TO$ $EG \to CO_2$ $EG \to EC$
Altıntaş (2013)	Turkey	EG, EC, I	Bounds/ Johansen-Juselius Coint., VECM Granger Causality Test	Yes	$EG \to CO_2$ $EC \to CO_2$ $I \to EG$
Aytun & Akın (2015)	Turkey	EG, EC, ED	Bootstrap Causality Test	-	$ED \to CO_2$ $ED \to EC$ $ED \to EG$
Bozkurt & Akan (2014)	Turkey	EG, EC	Johansen-Juselius Coint., Impulse-Response Analysis	Yes	Positive response from EC and EG to CO_2.
Bozkurt & Okumuş (2015)	Turkey	EG, EC, POP, TO	Hatemi-J Coint. Test	Yes	EKC is valid.
Chang (2010)	China	EG, EC	Multivariate Coint., Granger Causality Test	Yes	$EG \to CO_2$
Cowan et al. (2014)	BRICS	EG, EC	Granger Causality Test	-	Differing Results
Çetin & Ecevit (2015)	19 Sub Saharan African Countries	EC, UR	Pedroni/Kao Co-integration, Granger Causality Test	Yes	$CO_2 \leftrightarrow EC$ $CO_2 \leftrightarrow UR$
Çetin & Seker (2014)	Turkey	EG, TO	Bounds Coint.	Yes	+ impact of EG and TO on CO_2

Author	Country	Variables	Method	EKC valid?	Results
Dinda & Coondoo (2006)	88 Countries	EG	IPS Coint. Granger Causality Test	Yes	EKC is not valid $CO_2 \leftrightarrow EG$
Grossman & Krueger (1995)	42 Countries	EG	GLS	-	EKC is valid
Halıcıoğlu (2009)	Turkey	EG, EC, TO	Bounds Coint., Granger Causality Test	Yes	EKC is not valid $CO_2 \leftrightarrow EC$ $CO_2 \leftrightarrow EG$
Hossain (2011)	Newly Industrialized Countries	EG, EC, TO, UR	Johansen Coint., Granger Causality Test	Yes	$EC \rightarrow CO_2$ $TO \rightarrow CO_2$ $EG \rightarrow EC$ $UR \rightarrow EG$ $TO \rightarrow UR$
Hossain (2012)	Japan	EG, EC, TO, UR	Bounds/ Johansen and Juselius Coint., Granger Causality Test	Yes	$TO \rightarrow CO_2$ $EC \rightarrow CO_2$ $TO \rightarrow EC$ $CO_2 \rightarrow EG$ $EG \rightarrow TO$
Jalil & Mahmud (2009)	China	EG, EC, TO	Granger Causality Test	-	EKC is valid $EG \leftrightarrow CO_2$
Kapusuzoğlu (2014)	WW, OECD, EU Turkey	EG	Johansen Coint., Granger Causality Test	No (Turkey/EU) Yes (WW/OECD)	Differing Results
Kasman & Duman (2015)	New EU member and candidate countries	EG, EC, TO, UR	Pedroni/Kao/ Westerlund Coint., Granger Causality Test	Yes	EKC is valid $EC \rightarrow CO_2$ $UR \rightarrow CO_2$ $TO \rightarrow CO_2$ $UR \rightarrow TO\ EG \rightarrow EC$ $EG \rightarrow TO$ $EC \rightarrow TO$ $UR \rightarrow EG$

(continued on next page)

Tab. 2: (continued)

Authors	Country	Indicators	Model	Co-integration	Result
Katırcıoğlu (2017)	Turkey	EG, EC, OP	Maki Coint.	Yes	EKC is not valid
Katırcıoğlu et al. (2017)	Turkey	EG, EC, RER, M	Bounds Coint., Granger Causality Test	Yes	RER→EG
Katırcıoğlu & Celebi (2018)	Turkey	EG, EC, EXD	Maki Coint.	Yes	EKC is not valid. EG→EC CO_2→EXD EC→EXD
Katırcıoğlu & Taspinar (2017)	Turkey	EG, EC, FD	Maki Coint., Granger Causality Test	Yes	EKC is not valid
Katırcıoğlu & Katırcıoğlu (2018)	Turkey	EG, EC, FP	Maki Coint., Granger Causality Test	Yes	EKC is valid EG→CO_2 FP→CO_2 EC→CO_2
Keskingöz & Karamelikli (2015)	Turkey	EG, EC, TO	Bounds Coint.	Yes	TO, EC and EG increases CO_2
Kivyiro & Arminen (2014)	Six Sub Saharan Africa	EG, EC, TO	ARDL Coint, Granger Causality Test	Yes	Differing Results
Koçak (2014)	Turkey	EG, EC	Bounds Coint.	Yes	EKC is not valid
Lean & Smyth (2010)	ASEAN	EG, EC	Johansen Fisher Coint, Granger Causality Test	Yes	EKC is valid EC→CO_2

Study	Sample	Variables	Method		Differing Results
Lee & Lee (2009)	109 Countries	EG	Panel Seemingly Unrelated Regressions Augmented Dickey–Fuller	-	
Mohapatra & Giri (2015)	India	EG, EC, GFCF, UR, TO	Bounds Coint., Granger Causality Test	Yes	$EC \to CO_2$ $EG \to CO_2$ $TO \to CO_2$ $UR \to CO_2$ $EC \leftrightarrow EG$
Narayan & Smyth (2008)	G7	EG, EC, CAP	Pedroni/ Westerlund Coint.,	Yes	$EC \to EG$ $CAP \to EG$
Ozatac et al. (2017)	Turkey	EG, EC, UR, FD, TO	Bounds Coint., Granger Causality Test	Yes	EKC is valid. $EG \to CO_2$ $EC \to CO_2$ $UR \to CO_2$ $TO \to CO_2$
Öztürk & Acaravcı (2010)	Turkey	EG, EC, ER	Bounds Coint., Granger Causality Test	Yes	EKC is not valid. $ER \to EG$
Öztürk & Öz (2016)	Turkey	EG, EC, FDI	Bounds Coint., Granger Causality Test	Yes	$EC \to EG$ $CO_2 \to EC$
Pata (2018)	Turkey	EG, EC, FD, UR	Bounds Coint.	Yes	EKC is valid
Salahuddin et al. (2015)	GCC Countries	EG, EC, FD	Pedrioni Coint., VECM Granger Causality	Yes	$CO_2 \leftrightarrow EG$ $EC \to CO_2$

(continued on next page)

Tab. 2: (continued)

Authors	Country	Indicators	Model	Co-integration	Result
Tang & Tan (2015)	Vietnam	EG, EC, FDI	Johansen Coint., Granger Causality Test	Yes	EKC is valid $CO_2 \leftrightarrow EG$ $CO_2 \leftrightarrow FDI$
Yuan (2007)	China	EG, EC	Johansen Maximum likelihood Coint.	Yes	EC→EG
Yavuz (2014)	Turkey	EG, EC	Johansen/ Gregory and Hansen Coint. Tests	Yes	EKC is valid

ASEAN: Association of Southeast Asian Nations
BRICS: Brasil, Russia, India, China, South Africa
CAP: Capital Formation
CO_2: Carbon dioxide Emissions
EC: Energy Consumption
ED: Education
EG: Economic Growth
ER: Employment Ratio
EU: European Union
FD: Financial Development
FDI: Foreign Direct Investment
FP: Fiscal Policy
GCC: Gulf Cooperation Council
GFCF: Gross Fixed Capital Formation
I: Investment
M: Imports
OECD: Organization for Economic Cooperation and Development
OP: Oil Prices
POP: Population
RER: Real Exchange Rate
TO: Trade Openness
UR: Urbanization
WW: Worldwide

to be long enough to let "the ecosystems to adapt naturally to climate change, to ensure that food production is not threatened and to enable economic development in a sustainable manner" (United Nations, 1992). The treaty consists of 26 articles outlining the framework to achieve the objectives and 2 annexes as Annex I and Annex II listing the countries separated for different obligations depending on their economic development levels and responsibilities. Annex 1 consists of 43 Parties including industrialized countries and countries characterized as economies in transition, while Annex 2 countries are Organization for Economic Cooperation and Development (OECD) and EU members. Annex 1 includes countries that are obliged to improve policies for the prevention of GHG, while Annex II countries are responsible to guide developing countries and economies in transition to reduce their GHG emissions through providing financial and technical support for them. There is no binding limit on GHG emission generation levels set for individual countries. Furthermore, no country is faced with an enforcement mechanism. The yearly meetings are held since 1995 in Conferences of the Parties (COP) (United Nations, 1992).

UNFCCC is extended with the Kyoto Protocol that obliged parties to reduce GHG emissions by setting internationally binding emission reduction targets relying on the scientific consensus stating that the main contributor to global warming is human-made CO_2 emissions. The Protocol is accepted as being the first step to create a global emission reduction regime. It includes 192 parties and adopted in 1997 entering into force in 2005. Kyoto Protocol places a heavier burden on developed countries since it is recognized that the dominant reason behind the GHG emissions – hence global warming – is high levels of industrial activity for more than 150 years. There are 2 commitment periods of the Protocol; first started in 2008 and ended in 2012 and second agreed on in 2012, known as the Doha Amendment to the Kyoto Protocol. The countries obliged to reduce GHG emissions are listed in Annex B of the Protocol (United Nations, 1998).

Targets of the Protocol which should be met primarily through national measures of each country are listed in Annex B. However, there are three different market based mechanisms provided by the protocol, offering those countries an additional means to meet their targets. The main aim of those mechanisms is to help Annex B countries to meet their emission targets through stimulated green investment in a cost-efficient way. Annex B countries have accepted GHG emissions reduction targets on average 5 % below their 1990 levels for the first commitment period of the protocol from 2008 to 2012. Reduction targets are expressed as levels of allowed emissions which are divided to assigned amount units (AAUs).

1   Emissions Trading (ET) allows Annex B countries to trade AAUs so that countries not generating as much as the permitted level of emissions can trade AAUs with countries exceeding their targets. AAUs can be accepted as a new commodity to be traded creating the "carbon market", since CO_2 emission is the key contributor of global warming. Furthermore, there are several carbon markets created for trade purposes by some of the Annex B countries, such as EU Emissions Trading System (EU-ETS), Japan Emissions Trading System, Norway Emissions Trading System etc. (Öztürk et al., 2012).
2   Clean Development Mechanism (CDM) is the second mechanism. Since it does not matter for the climate where GHG emissions reduction takes place, CDM aims to achieve the emission reduction wherever it can be achieved with the lowest cost. CDM creates the opportunity for the Annex B countries to contribute to their targets while promoting the assistance of achieving sustainable development for developing countries with insufficient technology and flexible environmental policy through environmentally friendly investment from Annex B countries. Since the projects are cheaper to implement in developing countries, Annex B country governments or private businesses apply for projects for emission reduction in developing countries and receive certified emission reductions (CERs) which is counted against their reduction targets as subtracted from their emission reduction units.
3   Third mechanism is the Joint implementation (JI). JI mechanism allows any Annex B country to invest in any project in another Annex B country where it is cheaper to reduce emissions. To do so, the investing country is to gain emission reduction units (ERUs) which are counted to meet their targets.

The second commitment period of the Protocol was agreed in 2012 which binds 37 countries of Annex B. The key features of this period is that some of the countries included in Annex B have agreed to make further emission cuts, new rules were imposed for some countries of Annex B about how to account for emission from land use and forestry and nitrogen trifluoride (NF3) is included to the list of GHGs.

The Paris Agreement is a separate instrument under UNFCCC signed in December 2015. It is known to be the first-ever universal legally binding global climate deal, signed by 195 countries, which also sheds light and brings the effort for assisting developing countries to combat climate change. The main aim of the Agreement is to strengthen the global response to control climate change. To do so, an action plan is set to decrease the level of global warming below 2^0C above pre-industrial levels to be able to reduce the effects of climate change by the end of the century. Under the Paris Agreement, each country is responsible to submit

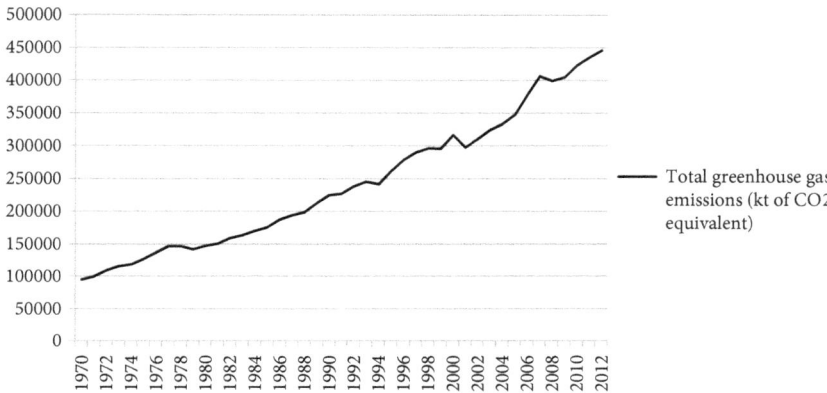

Fig. 2: GHG Emissions in Turkey. Source: Worldbank Database, 2018, https://data.worldbank.org/indicator/EN.ATM.GHGT.KT.CE?locations=TR&view=chart.

a national climate action plan called Nationally Determined Contributions (NDCs) which does not force any country to set a specific target by a specific date but involves targets set beyond the previously set targets. Paris Agreement came into action in 2016.

3 Environmental Degradation and Projects to GHG Generation in Turkey

Turkey is a developing country with high levels of energy use with very low shares of renewable energy sources, as illustrated by Fig. 1 in the first part of the chapter, accompanied with an increasing trend of GHG emissions which is illustrated in Fig. 2 below.

As it can be seen in Fig. 2, since 1970 there have been a constant increase in GHGs in Turkey which is accelerated after 1990s with the major economic changes leading to rapid economic growth and structural reformation of the economy through privatization of the state enterprises, price liberalization and integration in the European and global economy. Fig. 3 below illustrates the sectoral distribution of GHG generation in 2016.

As it is clear in Fig. 3, the energy sector shares 73 % of the GHGs generation in Turkey, being the dominant contributor to the increase in global warming and thus climate change. The reason behind the high level of energy sector GHG generation depends on the low share of renewable energy sources use.

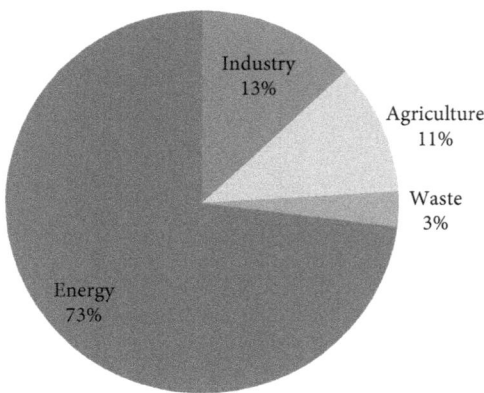

Fig. 3: GHGs Generated by Sectors, 2016. Source: TUIK, 2018, http://www.tuik.gov.tr/basinOdasi/haberler/2018_12_20180430.pdf

GHGs emission generation in Turkey generated by energy sector increased from 134.4 million tons to 340 million tons over the years between 1990 and 2015.

The highest share of GHG belongs to CO_2 emissions among other GHGs according to 2016 records. The share of CO_2 emissions generated accounts for more than 80 % of the total GHGs. Fig. 4 illustrates the share of GHGs generated in Turkey in 2016.

Since the share of CO_2 emissions in total GHGs is the highest at a significant rate, it requires detailed focus. CO_2 emissions occur mostly as a result of fuel consumption in the form of either gaseous, liquid or solid. Fig. 5 shows the trends in total CO_2 emissions, CO_2 emissions from gaseous fuel consumption, CO_2 emissions from liquid fuel consumption and CO_2 emissions from solid fuel consumption in Turkey over the years between 1978 and 2014.

There is an increasing trend in total CO_2 emissions, and also in CO_2 emissions generated from each type of fuel consumption separately. But it is clear that the highest increase in CO_2 emissions appears due to consumption of gaseous fuel over the last couple of years, but still the highest CO_2 emissions is due to solid fuel consumption.

Environmental concerns in Turkey gained importance at the end of 1970s. Government put some national efforts for the central organization of environmental management between the years 1978 and 2003 through five different steps; starting in 1978 with the settlement of Under Secretariat of Environment attached to the Prime Ministry, followed by the establishment of General

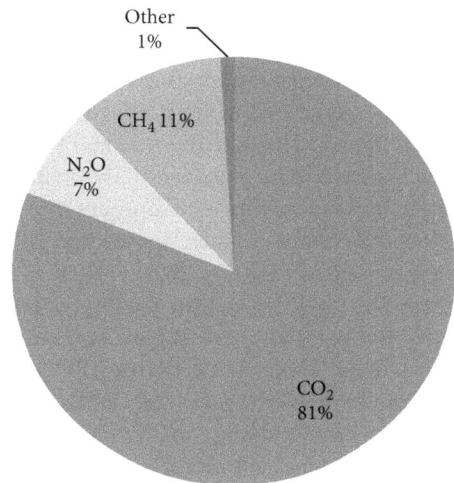

Fig. 4: Share of GHGs in Turkey in 2016. Source: TUIK, 2018, http://www.tuik.gov.tr/basinOdasi/haberler/2018_12_20180430.pdf.

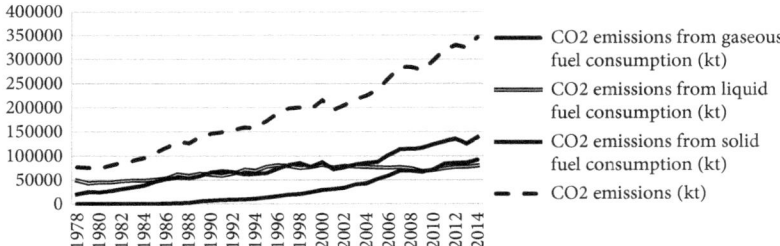

Fig. 5: CO_2 Emissions in Turkey. Source: Worldbank Database, 2018, https://data.worldbank.org/indicator/EN.ATM.GHGT.KT.CE?locations=TR&view=chart.

Directorate of Environment in 1984, Under Secretariat of Environment in 1989, Ministry of Environment in 1991 and finally Ministry of Environment and Forestry in 2003.

Being an OECD member, when UNFCCC was adopted, Turkey was listed under both Annex I and Annex II countries. This position is rejected by Turkey realizing the consequences of this classification. Turkey put diplomatic efforts to change this situation and in 2001 she got removed from Annex II. So having a special status Turkey is listed only in Annex I since 2002.

Turkey officially gained candidate status for the European Union (EU) in 1999. To become a member country of EU Turkey has to fulfill the acquis communautaire which is the EU legal system consisting of 35 chapters. Chapter 27 of the aquis, opened in 2009, is related to environmental issues. Turkey's eighth development plan was designed to include the National Environmental Strategy and Action Plan to tackle environmental issues. The main target is to ensure sustainable development while protecting environment through preventing pollution and protecting natural resources. There are several projects put into action to achieve the EU environmental target in Turkey; such as Implementation of Nitrate Directive, Institution Building on Air Quality in the Marmara Region Environment, Capacity Building in the Field of Environment, The REACH Chemicals Project, Improving Emissions Control, IPPC-Integrated Pollution Prevention Control, Mining Waste Management, Control of Industrial Organic Volatile Compounds, Capacity Building on Water Quality Monitoring etc. (Republic of Turkey Ministry of Development, 2017).

In line with the opening of chapter 27, Turkey became a party to the Kyoto protocol in the same year. Since Turkey is not listed under Annex II, there is no any legal emission reduction commitments for Turkey. However, according to the information obtained from OECD DAC statistics of 2014, Turkey is one of the top ten countries of the bilateral climate finance receiving. The bilateral climate finance received in 2015 is even more than the value received in 2014. Furthermore, Turkey also received substantial funding from European institutions, such as the European Bank for Reconstruction and Development (ERBD) and the European Investment Bank (EIB). However, even though the energy use of Turkey is lowest among the EU members, she is the third highest CO_2 emission generating country following Romania and Austria. Turkey needs to reshape its national environmental policies, especially policies related to promoting renewable energy use to decrease CO_2 emissions.

Conclusion

Environmental issues started to gain more attention in both developed and developing countries in the recent years since the human health is highly affected from environmental conditions. As a result, consumers' demand for a clean environment increased and developed countries as well as most of the developing countries started to apply legislative regulations on the environment.

The most important concern related to environment is the problem of global warming which occurs mostly because of the GHGs generated during production. Many people are still not being informed enough related to the issue in most developing countries. Therefore, this study aims to give brief information about what causes global warming, what has been done in the international academic literature to find out what causes GHGs – in particular CO_2 emissions since CO_2 is the dominant GHG causing global warming –, the intergovernmental and binding agreements to control environmental degradation and effects and the situation in Turkey by giving information about environmental degradation and projects to control GHG generation in Turkey.

According to the literature related to the topic, as production increases, energy consumption raises, CO_2 emissions also ascends in Turkey. The dominant contributor of the increase in CO_2 emissions in Turkey is the high share of fossil fuels use for energy generation among the energy resources used both for production and consumption. Energy policy of Turkey is critically important not only considering the environmental problems caused by CO_2 emissions but also in terms of economic conditions; since energy production in Turkey mostly depends on imported resources from abroad, increasing cost of production and leading to the current account deficit. The share of renewable energy production in total energy generation can be increased; since she is a favorable country for renewable energy sources as mentioned before.

There are two main reasons that can explain the high levels of CO_2 emissions generations in Turkey. The first is related to the high level of energy intensity whereas the second is related to the low share of the renewable energy use in Turkey. Energy intensity is given by the amount of energy per unit of output produced. High energy intensity indicates relatively high levels of energy use to produce per unit output, which in other words is a sign for inefficient energy use for production of output. Energy intensity in Turkey is high due to the inefficient use of energy sources (İslatince & Haydaroğlu, 2009). Energy efficiency of Turkey is below the world average while energy intensity values are higher than the average of OECD countries. CO_2 emissions in Turkey can be decreased through decreasing energy intensity by increasing the share of renewable energy sources used for energy generation. One thing that can be done is to increase the carbon tax for energy production. Furthermore, financial assistance for research and development can be provided by the collected carbon taxes for technological innovations which might decrease energy intensity.

Turkey is a gifted country with renewable energy sources. However, renewable energy construction is costly and financing renewable energy production in the short run might be difficult. On the contrary, in the long term, increasing the share of renewable energy resources decreases dependency on imported energy decreasing the cost of production and promoting economic growth. Furthermore, some energy resources such as hydroelectric power have lower operational costs in Turkey having geographical advantages. Moreover, climate characteristics of Turkey provide good opportunities for wind and solar energy potentials having similar features with hydroelectric energy generation. Marmara, Aegean and Eastern Mediterranean regions offer good potential for wind energy generation, while Southeast Anatolia, Mediterranean and Eastern Anatolia regions have high potentials for solar energy production (Yılmaz, 2012).

Turkey should achieve sustainable economic growth accompanied with reshaped energy policies that promote the use of renewable energy to decrease its dependency on foreign energy resources and to control the environmental degradation through decreasing GHGs generation, especially CO_2 emissions.

References

Acaravci, A., and Ozturk, I. (2010). "On the Relationship between Energy Consumption, CO_2 Emissions and Economic Growth in Europe", Energy, 35(12): 5412–5420.

Akbostancı, E., Türüt-Aşık, S., and Tunç, G. İ. (2009) "The Relationship between Income and Environment in Turkey: Is there an Environmental Kuznets Curve?", Energy Policy, 37(3): 861–867.

Akin, C. S. (2014) "The Impact of Foreign Trade, Energy Consumption and Income on CO_2 Emissions", International Journal of Energy Economics and Policy, 4(3): 465.

Altıntaş, H. (2013). "Türkiye'de Birincil Enerji Tüketimi, Karbondioksit Emisyonu ve Ekonomik Büyüme İlişkisi: Eşbütünleşme ve Nedensellik Analizi", Eskişehir Osmangazi Üniversitesi İktisadi ve İdari Bilimler Dergisi, 8(1): 263–294.

Aytun, C., and Akın, C. S. (2015) "Türkiye'de Karbondioksit Emisyonu, Enerji Tüketimi ve Eğitim İlişkisi: Bootstrap Nedensellik Analizi", Düzenleme Kurulu, 260–273.

Bilgen, S., Keleş, S., Kaygusuz, A., Sarı, A., and Kaygusuz, K. (2008) "Global Warming and Renewable Energy Sources for Sustainable Development: A Case Study in Turkey", Renewable and Sustainable Energy Reviews, 12(2): 372–396.

Bozkurt, C., and Akan, Y. (2014) "Economic Growth, CO_2 Emissions and Energy Consumption: The Turkish Case", International Journal of Energy Economics and Policy, 4(3): 484.

Bozkurt, C., and Okumuş, İ. (2015) "Türkiye'de Ekonomik Büyüme, Enerji Tüketimi, Ticari Serbestleşme ve Nüfus Yoğunluğunun CO_2 Emisyonu Üzerindeki Etkileri: Yapısal Kırılmalı Eşbütünleşme". Mustafa Kemal Üniversitesi Sosyal Bilimler Enstitüsü Dergisi, 12(32): 23–35.

Center for Climate and Energy Solutions (2017) https://www.c2es.org/facts-figures/main-ghgs Access Date: 17.07.2018

Chang, C. C. (2010) "A Multivariate Causality Test of Carbon Dioxide Emissions, Energy Consumption and Economic Growth in China", Applied Energy, 87(11): 3533–3537.

Çetin, M., and Seker, F. (2014) "Ekonomik Büyüme ve Dış Ticaretin Çevre Kirliliği Üzerindeki Etkisi: Türkiye İçin Bir ARDL Sınır Testi Yaklaşımı. Yönetim, 21(2):213–230.

Çetin, M., and Ecevit, E. (2015) "Urbanization, Energy Consumption and CO_2 Emissions in Sub-Saharan Countries: A Panel Co-integration and Causality Analysis", Journal of Economics and Development Studies, 3(2): 2334–2390.

Cowan, W. N., Chang, T., Inglesi-Lotz, R., and Gupta, R. (2014) "The Nexus of Electricity Consumption, Economic Growth and CO_2 Emissions in the BRICS Countries", Energy Policy, 66: 359–368.

Dinda, S., and Coondoo, D., (2006) "Income and Emission: A Panel-Data Based Co-Integration Analysis", Ecological Economics 57: 167–181.

Grossman, G.M., and Krueger, A.B., (1995) "Economic Growth and the Environment", Quarterly Journal of Economics 110: 353–377.

Halicioglu, F. (2009) "An Econometric Study of CO_2 Emissions, Energy Consumption, Income and Foreign Trade in Turkey", Energy Policy, 37(3): 1156–1164.

Hossain, M. S. (2011). "Panel Estimation for CO_2 Emissions, Energy Consumption, Economic Growth, Trade Openness and Urbanization of Newly Industrialized Countries", Energy Policy, 39(11): 6991–6999.

Hossain, S. (2012). "An Econometric Analysis for CO_2 Emissions, Energy Consumption, Economic Growth, Foreign Trade and Urbanization of Japan", Low Carbon Economy, 3(3):92–105.

International Energy Agency, 2017, https://www.iea.org/ Access Date: 10.07.2018

Islatince, H., and Haydaroğlu, C. (2009) "Türk İmalat Sanayiinde Enerji Verimliliği ve Yoğunluğunun Analizi", Dumlupınar Üniversitesi Sosyal Bilimler Dergisi, 24, Retrieved from http://dergipark.gov.tr/dpusbe/issue/4766/65525.

Jalil, A., and Mahmud, S. F. (2009) "Environment Kuznets Curve for CO_2 Emissions: A Cointegration Analysis for China", Energy Policy, 37(12): 5167–5172.

Kapusuzoğlu, A. (2014) "Causality Relationships between Carbon Dioxide Emissions and Economic Growth: Results from a Multi-Country Study", International Journal of Economic Perspectives, 6: 5–15.

Kasman, A., and Duman, Y. S. (2015). "CO_2 Emissions, Economic Growth, Energy Consumption, Trade and Urbanization in New EU Member and Candidate Countries: A Panel Data Analysis", Economic Modelling, 44: 97–103.

Katircioglu, S. (2017) "Investigating the Role of Oil Prices in the Conventional EKC Model: Evidence from Turkey", Asian Economic and Financial Review, 7(5): 498–508.

Katircioglu, S., Katircioğlu, S. and Altinay, M. (2017) "Interactions between Energy Consumption and Imports: Empirical Evidence from Turkey", Journal of Comparative Asian Development, 16 (2): 161–178.

Katircioglu, S., and Taspinar, N. (2017) "Testing the Moderating Role of Financial Development in an Environmental Kuznets Curve: Empirical Evidence from Turkey", Renewable and Sustainable Energy Reviews, 68: 572–586.

Katircioglu, S., and Celebi, A. (2018) "Testing the Role of External Debt in Environmental Degradation: Empirical Evidence from Turkey", Environmental Science and Pollution Research, 25(9): 8843–8852.

Katircioglu, S., and Katircioğlu, S. (2018) "Testing the Role of Fiscal Policy in the Environmental Degradation: The Case of Turkey", Environmental Science and Pollution Research, 25: 5616–5630.

Kesgingöz, H., and Karamelikli, H. (2015) "Dış Ticaret, Enerji Tüketimi ve Ekonomik Büyümenin CO_2 Emisyonu Üzerine Etkisi", Kastamonu University Journal of Economics & Administrative Sciences Faculty, 9: 7–17.

Kivyiro, P., and Arminen, H. (2014) "Carbon Dioxide Emissions, Energy Consumption, Economic Growth and Foreign Direct Investment: Causality Analysis for Sub-Saharan Africa", Energy, 74: 595–606.

Koçak, E. (2014) "Türkiye'de Çevresel Kuznets Eğrisi Hipotezinin Geçerliliği: ARDL Sınır Testi Yaklaşımı", İşletme ve İktisat Çalışmaları Dergisi, 2(3): 62–73.

Kraft, J. and Kraft, A. (1978) "On the Relationship between Energy and GNP", Journal of Energy and Development, 3(2): 401–3.

Lean, H. H. and Smyth, R. (2010) "CO_2 Emissions, Electricity Consumption and Output in ASEAN", Applied Energy, 87(6): 1858–1864.

Lee, C.C. Lee, J.D (2009) "Income and CO_2 Emissions: Evidence from Panel Unit Root and Co-Integration Tests", Energy Policy, 37: 13–423.

Mohapatra, G., Giri, A. K., (2015)" Energy Consumption, Economic Growth and CO_2 Emissions: Empirical Evidence from India", The Empirical Econometrics and Quantitative Economics Letters, 4(1): 17–32.

Narayan, P. K. and Smyth, R. (2008) "Energy Consumption and Real GDP in G7 Countries: New Evidence from Panel Co-Integration with Structural Breaks", Energy Economics, 30(5): 2331–2341.

Onat, A., İmal, M. and İnan, A. T. (2004) "Soğutucu Akışkanların Ozon Tabakası Üzerine Etkilerinin Araştırılması ve Alternatif Soğutucu Akışkanlar", KSÜ Fen ve Mühendislik Dergisi, 7(1): 32–38.

Ozatac, N., Gokmenoglu, K. K. and Taspinar, N. (2017) "Testing the EKC Hypothesis by Considering Trade Openness, Urbanization and Financial Development: The Case of Turkey", Environmental Science and Pollution Research, 24: 16690–16701.

Özmen, T. (2009) "Sera Gazı, Küresel Isınma ve Kyoto Protokolü", İMO Dergisi, 453(1): 42–46.

Öztürk, I. and Acaravci, A. (2010) "CO2 Emissions, Energy Consumption and Economic Growth in Turkey", Renewable and Sustainable Energy Reviews, 14(9): 3220–3225.

Öztürk, A., Demirci, U. and Türker, M. F. (2012) "İklim Değişikliğiyle Mücadelede Karbon Piyasaları ve Türkiye için bir Değerlendirme", Ulusal Akdeniz Orman ve Çevre Sempozyumu, Kahramanmaraş, Special Edition: 306–312.

Öztürk, Z. and Öz, D. (2016) "The Relationship between Energy Consumption, Income, Foreign Direct Investment and CO2 Emissions: The Case of Turkey", Çankırı Karatekin Üniversitesi Journal of Faculty of Economics and Administrative Sciences, 6(2):269–288.

Papadimitriou, V. (2004) "Prospective Primary Teachers' Understanding of Climate Change, Greenhouse Effect and Ozone Layer Depletion", Journal of Science, Education and Technology, 13(2): 299–307.

Pata, U. K. (2018) "Renewable Energy Consumption, Urbanization, Financial Development, Income and CO_2 Emissions in Turkey: Testing EKC Hypothesis with Structural Breaks", Journal of Cleaner Production, 187: 770–779.

Republic of Turkey Ministry of Development, 2017, "Development Plans", http://www.surdurulebilirkalkinma.gov.tr/wp-content/uploads/2016/07/2030_Raporu.pdf Access Date: 10.01.2019.

Salahuddin, M. Gow, J. ve Ozturk, I. (2015) "Is the Long-Run Relationship between Economic Growth, Electricity Consumption, Carbon Dioxide Emissions and Financial Development in Gulf Cooperation Council Countries Robust?", Renewable and Sustainable Energy Reviews, 51: 317–326.

Tang, C. F. and Tan, B. W. (2015). "The Impact of Energy Consumption, Income and Foreign Direct Investment on Carbon dioxide Emissions in Vietnam", Energy, 79: 447–454.

TUIK, (2018), http://www.tuik.gov.tr/basinOdasi/haberler/2018_12_20180430.pdf Access Date:12.07.2018

Turkish Statistical Institute, 2018, http://www.turkstat.gov.tr/ Access Date: 02.07.2018

United Nations, 1992, "United Nations Framework Convention on Climate Change" https://unfccc.int/resource/docs/convkp/conveng.pdf Access Date: 05.08.2018

United Nations, 1998, "Kyoto Protocol to the United Nations Framework Convention on Climate Change", https://unfccc.int/resource/docs/convkp/kpeng.pdf Access Date: 17.07.2018

World Bank, 2018 "World Development Indicators", https://data.worldbank.org/indicator/EN.ATM.GHGT.KT.CE?locations=TR&view=chart. Access Date: 21.07.2018

Yavuz, N. Ç. (2014) "CO_2 Emissions, Energy Consumption and Economic Growth for Turkey: Evidence from a Co-Integration Test with a Structural Break", Energy Sources, Part B: Economics, Planning, and Policy, 9(3): 229–235.

Yılmaz, M. (2012) "Türkiye'nin Enerji Potansiyeli ve Yenilenebilir Enerji Kaynaklarının Elektrik Enerjisi Üretimi Açısından Önemi", Ankara Üniversitesi Çevrebilimleri Dergisi, 4(2): 33–54.

Yuan, J.H., Zhao, C.H., Yu, S.K., and Hu, Z.G. (2007) "Electricity Consumption and Economic Growth in China: Co-Integration and Co-Feature Analysis", Energy Economics 29: 1179–1191.

Tunç Durmaz

Environmental Catastrophes and Deep Uncertainties Surrounding the Economics of Climate Change

Introduction

There is a big debate in climate change which is a natural result of small knowns and big unknowns. While one party demands immediate action by looking at the small knowns, the other party demonstrates a "wait-and-see" policy. The problem is that the benefits from mitigation activities are distant and uncertain but the investment costs are immediate and high, and extends to a longer period of time.

Evaluating benefits and costs necessitates an economist to use discounting in the analysis. However, since the climate change is a long-term problem and there are deep uncertainties regarding its consequences, a usual integrated assessment model (IAM) to conduct a cost-benefit analysis (CBA) might be an oversimplification of the future state that we will have to face. As a result, the question is how to incorporate structural uncertainty into the analysis.

This chapter focuses on different perspectives about discounting. Firstly, in Section 1, we look at discounting under certainty. Such simplicity will be helpful when we consider discounting under structural uncertainty. In Section 2, we focus on the structural uncertainty. This is followed by a discussion of the Dismal Theorem in Section 3. In Section 4, we examine the structural uncertainties in the functional forms of utility. Lastly, Section 5 concludes.

1 Discounting under Certainty in a Single World

The economics of climate change is frequently outlined as a choice among two scenarios. In one scenario we can prevent future losses by taking costly actions today, but this would imply cuts in GDP in the short run. In the other situation, we can decide to live in a world where we do nothing but eventually will suffer severe income losses from the damages coming from global warming (Conceição and Zhang, 2010). This choice among two scenarios suggests trade-offs across generations because climate change is a phenomenon that takes place in the long run. Although it is relatively a simple concept, the discount rate discussion is at the heart of many debates surrounding global warming. Some questions in this regard are: What is the level of importance that we should assign to the welfare of generations of the future versus the current ones?; Will the generations in the future be richer because of sustained economic growth?; What will be the burden for the poor and the rich?

We know as well as feel that the earth is warming and more than half of the observed increase in global average surface temperature since more than fifty years was caused by the anthropogenic increase in GHG concentrations and other anthropogenic forcings together (IPCC, 2014). Nevertheless, knowing the pace of the temperature and its future implications with certainty is difficult.

Catastrophic outcomes can alter Earth fundamentally. The dynamic nature of climate change cannot be adequately analyzed in a deterministic setting. Hence it is essential to incorporate risk and uncertainty into the analysis. To understand discounting and its implications in a dynamic context, first, we will look at discounting under certainty. Such simplicity will be helpful when we consider discounting under structural uncertainty.

Let $u(c_t)$ denote the utility at time t where c_t is consumption. Assume that the marginal utility is positive but declines with consumption; that is, $u'(c_t) \equiv \frac{\partial u(c_t)}{\partial c_t} > 0$ and $u''(c_t) \equiv \frac{\partial^2 u(c_t)}{\partial c_t^2} < 0$, respectively. If δ is the discount rate, intergenerational welfare is the discounted sum of generational utilities, that is, discounted sum of $u(c_t)$s

$$\sum_{t=0}^{\infty} \frac{1}{(1+\delta)^t} u(c_t). \tag{1}$$

Let ρ stand for the consumption discount rate (social rate of discount). Therefore, a unit of consumption at time t is substituted for $1 + \rho$ units of consumption at time $t + 1$. Suppose that a benevolent planner is provided a consumption trajectory for c_t. Furthermore, let dc_t and dc_{t+1} denote variations in consumption which does not alter the total welfare given by Eq. (1). Then,

$$\frac{u'(c_t)}{(1+\delta)^t} dc_t + \frac{u'(c_{t+1})}{(1+\delta)^{t+1}} dc_{t+1} = 0. \tag{2}$$

Assuming a constant relative risk aversion (CRRA) utility function,

$$u(c_t) = \begin{cases} c_t^{1-\eta}/(1-\eta), & \text{for } \eta > 0 \text{ and } \eta \neq 1, \\ \ln(c_t), & \text{for } \eta = 0, \end{cases} \tag{3}$$

where η is the elasticity of the marginal utility, and from

$$-dc_t = \frac{dc_{t+1}}{1+\rho}, \qquad (4)$$

Eq. (2) can be rewritten to give

$$1+\rho = (1+\delta)(1+g(c_t))^\eta. \qquad (5)$$

where $g(c_t) \equiv c_{t+1}/c_t - 1$ is consumption growth. For any small number x, $\ln(1+x) \approx x$. Consequently, Eq. (5) reduces to

$$\rho_t \approx \delta + \eta g(c_t). \qquad (6)$$

In the optimum, the rate of return on capital is decided by the economic growth rate, aversion to consumption inequality among generations and generational time preference. Ceteris paribus, a higher discount rate (δ) results in a higher ρ_t. The larger is η, ceteris paribus, the larger ρ_t gets if the consumption growth is positive ($g(c_t)>0$). Thus, η is the aversion index the society displays toward consumption inequality among individuals as well as generations (Dasgupta, 2008).

When it comes to how much and how fast we should react to the threat of global warming, Nordhaus (2007) favors policies that are aimed at slowing climate change increasingly constrict emissions over time. He adds that in the following decades when damages are predicted to rise relative to output, it will become efficient to move investments toward intensive abatement. The exact mix and timing of emission reductions depend on the details of damages, costs and the nonlinearity and irreversibility of climate change and its damages.[1] As a result, δ and η should be calibrated to match the market interest rates, observed values of rate of consumption growth, $g(c_t)$ and private and public saving and investment rates.

According to Dasgupta (2008), this is an appealing and democratic move, but it seems that there is serious a difficulty when the concern is world's climate. Dasgupta (2008) explains that there are two unknowns that Nordhaus must determine (which δ and η), but only one equation, $\rho_t = r$, where r is the money market interest rate that relates them. This leads to the problem of consistency. Moreover, climate change under business-as-usual (BAU) involves a substantial global commons problem. He adds that the social rates of return on investments leading to energy-intensive activities are negative today, but the market does

---

1 The last sentence may be seen as inconsistent since irreversibility and nonlinearity might not imply a sufficient recovery, making climate-policy ramp impracticable at the first sight.

not reflect this because private rates would necessarily be favorable. There is a good reason to think that the observed values of consumption growth rates are not the ones the society would prefer were they able to choose collectively from the considerations presented earlier. Hence, there is a serious chance that the observed behavior offers a wrong basis for calibrating δ and η.

However, by relying exclusively on the revealed preference of an individual, Nordhaus has been consistent while Stern has not (Dasgupta, 2008). In the review, Stern chose η using estimates obtained from consumer behavior. When it comes to determining δ, however, Stern overlooks the consumers and looks for the advice of moral philosophers.

Nordhaus (2007) claims that discounting requires us to look carefully at the real interest rate as the benchmark for climatic investments. Therefore, the normatively acceptable and low-interest rates are irrelevant for determining the appropriate discount rate and use in the financial and capital markets. Countries will look at the actual gains from deals, and the returns on these investments relative to others, rather than the benefits that would arise from a theoretical growth model (Nordhaus, 2007). However, in this case, the question would be about the appropriateness of putting the climate change issue into a bargaining context when the globe as a whole will be affected in the future. Therefore, the discount rate should not necessarily be the same as the market interest rates.

Another intriguing argument in Nordhaus (2007) is that the Review's strategy would worsen the welfare of future generations, and therefore, would be Pareto deteriorating. The reason would be too many investments in low-yield abatement strategies which would be made too early. He claims that after five decades, conventional capital would be much lessened, while climate capital would only be somewhat increased. The best strategy is to invest more in the conventional capital today and use the additional resources in the future to invest in climate capital. Yet, one can argue that Nordhaus overlooks the possible irreversibility of climate change, and accordingly, structural uncertainty. More emphasis will be given on this issue in the following parts, in particular when we look at the debate between Martin Weitzman and William D. Nordhaus over the Dismal Theorem.

One crucial point in Nordhaus' analysis is his argument that the fundamental distinction between the Stern Review and other studies is the implicit return on capital. Precisely at this point, it is undeniable that the Report is political.

Sterner and Persson (2008) argue that using a market rate as a benchmark cannot be suitable because of ethical or normative judgments. The market is not merely observed as is done in positive or empirical studies. Instead, arguments for public action necessitate public goods that are difficult to provide. The Ramsey framework offers a setup to organize the views on this issue, and normally, it is essential to evaluate the rates in light of the observable rates in the market. However, the latter cannot be the only determinants of whether the values of δ and η are appropriately chosen. This is because it is pointless to attempt this exercise on ethical grounds, and what is normative would not be independent of what is positive. As Hume had put it long ago, one cannot derive an ought from an is: it is crucial that the phenomenon is observed and explained and a reason is provided (Hume, 1739). Therefore, the argument over the rate of discount is a problem for which the answer is closely involved with the value judgments.

The main point in Sterner and Persson (2008) is that one should consider the prosperity of future generations. If there is economic growth but deteriorating environmental quality, the relative value of the environmental goods need to be taken into consideration when evaluating future welfare gains and losses. There is no doubt that this analysis will deliver discount rates that are close to Stern's because of the environmental corrosion. However, I find their argument and analysis appealing and to the point.

Using DICE-2007, Sterner and Persson modify the utility function and include an additional equation that determines how the consumption of environmental goods changes over time in response to climate change.[2] In the DICE model utility function is of the CRRA form given by Eq. (3). To include the effect coming from the degrading climate, the utility function is modified in Sterner and Persson (2008) to take the following form:

$$u(c_t) = \frac{1}{1-\eta}\left[(1-\gamma)c_t^{1-1/\sigma} + e_t^{1-1/\sigma}\right]^{(1-\eta)\sigma/(\sigma-1)} \quad (7)$$

where e represents environmental amenities. The elasticity of substitution between consumption and environmental amenity is given by σ, η is the elasticity of marginal utility of consumption and $\gamma \in (0,1)$. It is further assumed that the consumption of environmental amenities is influenced by temperature rise only. Therefore, in the absence of temperature rise, and accordingly, climate change, environmental quality will neither worsen nor improve. The relationship between temperature and consumption of environmental amenities is given by

$$e_t = \frac{e_0}{1+aT_t^2} \quad (8)$$

---

[2] The reader is referred to Nordhaus (2016) for more information on the DICE model and recent projections.

where e_0 is the current level of environmental amenities, a is a positive constant and T_t stands for the temperature above the pre-Industrial Revolution (of pre-warming) level.

Sterner and Persson (2008) find that with $\sigma = .5$ and $\eta = 2$, temperature rise, and therefore, damages, enter the utility function additively alongside with consumption:

$$u(c_t, T_t) = -\left[\frac{1}{c_t} + (1 + \theta T_t^2)\right] \qquad (9)$$

where $\theta \equiv a\gamma / \left[(1-\gamma)e_0\right]$, which is calibrated to a predicted loss coming from a temperature rise of 2 to 3°C. Plugging Eq. (9) into the DICE model yields a far more stringent emissions policy than what Nordhaus calculates with a utility function in multiplication form:

$$u(c_t, T_t) = \left[-\frac{1}{c_t}(1 + \theta_N T_t^2)\right] \qquad (10)$$

where θ_N is a positive coefficient which is also calibrated to a loss coming from a 2 to 3°C rise in the global temperature.

2 Structural Uncertainty

Because of man-made GHG emissions worldwide, the average temperature in the world has been rising. No wonder that this is affecting the environment in general. However, we are unsure about the size and the timing of the impacts of a changing climate. Hence, we do not know with certainty when to act and how strong and significant this action should be. As a result, we are exposing ourselves to risks, but it is tough to tell what these risks are and what to do against them.

There is a big problem of commons and political parties, interest groups, government officials and so on, attach threats to various objects differently. For example, the ones who are most concerned about economic growth and development, for instance, have a tendency to be less concerned about environmental pollution or climate change.

As the importance level of different concerns can be ranked, so the dangers. However, this demands agreement criteria in advance, and no standardized procedure exists to rank these concerns. Specifically, risks are conceptually uncontrollable and even after a hazard has taken place, how much more action would have been necessary to have prevented it and whether such action would have been within the bounds of "reasonable" behavior would still be not clear. Therefore, we face a similar situation where numerous people are unsure whether enough is being done to prevent (extreme) climate change. There is a lack of knowledge as well as consent, and under these circumstances, it is almost impossible to come up with a solution (Douglas and Wildavsky, 1982, p.5).

Climate change is a complicated issue and involves various disciplines adding further difficulties to the problem. On the one hand, Weitzman has proposed Dismal Theorem which tells that the society has an indefinite utility loss from low-probability and high-impact catastrophes, and therefore, the price of future consumption is exceptionally high (infinite in the limit). Hence, to avoid such high prices, there is an immense need for a catastrophe insurance today. He adds that the policy advice coming from conventional CBA of climate change should be sifted through until low-probability and high impact events are incorporated in the study. Nordhaus challenges the Dismal Theorem and argues that it holds only under limited conditions. He adds that although Weitzman makes an essential point concerning the choice of probability distributions, the conditions needed for the theorem to hold are restricted and the range of potential scenarios that it can be applied to is narrow. As a result, the grounds that conventional IAMs can be discredited are limited, and therefore, they still have a strong-hold in climate change analysis.

In the following, the study will focus on the structural uncertainty that Weitzman (Weitzman, 2009b) has carried into the climate change analysis, evaluate his results and look at the Nordhaus critique (Nordhaus, 2009). The structural uncertainty in the functional form of utility (Weitzman, 2011) will be examined in the last part.

Weitzman (2007a) is mainly concerned with the amount of insurance to compensate a catastrophe with a small probability. He analyzes the interaction among risk, uncertainty and discounting, and the economics of climate change using a general equilibrium model. Beginning with the Ramsey equation given by Eq. (6), he takes g as a random variable (RV). The probability distribution of g has a left tail that is thickened due to climate change and contains a significant amount of the weight of expected marginal utility in CBA.

In Eq. (6) there is no differentiation among return rates on several assets, and ρ_t is the interest rate in the economy. However, when one considers a non-deterministic setting, there are various rates of return, and they differ considerably (think about different opinions individuals can have about the discount rates that they would use in discounting future payoffs of very-long-term investments). This can lead to tremendous differences depending on the discount rates employed. To better determine which discount rate to use, Weitzman

(2007a) argues that the Ramsey framework should be improved by incorporating uncertainty. This would enable one to distinguish between return rates on different kinds of investments.

Taking the growth rate to be a RV and supposing that g is independent and identically distributed (i.i.d.) normal with known mean μ and known variance σ^2 (that is, $g \sim N(\mu; \sigma^2)$), Eq. (6) becomes the generalized Ramsey formula

$$\rho^f \approx \delta + \mu\eta - \frac{1}{2}\eta^2\sigma^2,$$

where ρ_f is the risk-free interest rate.

Suppose that this model economy is represented by a dynamic stochastic general equilibrium as in the Lucas fruit-tree economy (Lucas, 1978). Let the RV P_e be the gross return on equity while $\rho^e \equiv \ln P^e$ is the geometric rate of return on equity. The equity risk premium over the risk-free rate then reduces to

$$\overline{\rho}^e - \rho^f = \eta\sigma^2$$

where $\overline{\rho}^e \equiv \ln \mathrm{E}\left[P^e\right]$. Combining the two previous equations yields the average return on equity

$$\overline{\rho}^e = \delta + \mu\eta - \frac{1}{2}\eta^2\sigma^2 + \eta\sigma^2.$$

Now the question turns out to be whether one should use the risk-free or the risky economy-wide return rate to discount costs and benefits of change in climate.[3] Weitzman (2007a) claims that while the theory predicts the average return on equity well, it fails when predicting the risk-free rate and the equity premium. This gives rise to the "risk-free rate" and the "equity premium" puzzles. Benefiting from numerical exercises, Weitzman (2007a) concludes that the emission reductions, as well as economic growth, that is subject to global warming, are very susceptible to the interest rate that is inherent in the model. He finds this resulting sensitivity somewhat discouraging.

One explanation of the asset-return puzzles, which can have relevance for the economics of climate change, is the idea that investors are overly scared of rare disasters. People are willing to pay high premiums for safer stores of value

---

3 A risky economy-wide rate of return applies to investments that have payoff characteristics which are parallel to the economy itself. On the contrary, a risk-free rate of return refers to investments whose payoffs are independent of the economy. Then the decision is to decide which of these two rates are more relevant for discounting costs and benefits of mitigating the effects of global warming.

that can represent "catastrophe insurance" (Weitzman, 2007a). Such an ongoing catastrophe-insurance can illustrate why the observed risk-free interest rate is very low when compared to the observed past average of return on equity. Then what makes these investors so alarmed that they look for a catastrophe insurance?

The Stern Review goes over these highly improbable but probably huge catastrophes associated with irreversible changes in the climate. Such irreversible changes include a sudden collapse of the ice sheets of Greenland and West Antarctica, weakening or even reversal of thermohaline circulations that can fundamentally affect Gulf Stream and climate in the European continent, rapid release of methane from the Arctic permafrost, crop failures and so on. Such rare disasters are all the way over in the right tail of very high temperature rise, T_t, corresponding to being far out in the left tail of $g(c_t)$.

The big debate about climate change is not only because of the big uncertainties about future temperature change and its consequences, but also there are significant opportunity costs of climate change projects (Schelling, 1995). These alternatives can include investments in public health, birth control, training and education, research, physical infrastructure and water resources. However, an important distinction here is that such alternative investments seem to compensate mostly for potential loss of indoor consumption and tend to be a lot expensive than wholesale abatement of GHGs.[4]

Weitzman's argument is such that the real problem is in the tails, more specifically, in the extremes with low probabilities but extreme effects, and mostly concerns "outdoor" consumption. If the definition of consumption would be expanded to incorporate the non-market use of the environment, such as habitats, ecosystems and species, then it is difficult to think of the compensating investments for which we should be saving as an alternative to keeping down the temperature rise, T_t.

Weitzman (2007a) sets aside the marginal analysis and let g stand for the uncertain consumption growth in the future including natural environments, ecosystems, species and the like. With a stochastic evolutionary process, such as climate change, the world may not persist for a sufficiently long period to acquire the knowledge to calculate the tail probabilities accurately. The net result is thicker left tails for the probability density function (PDF) of g.

---

4   Weitzman (2007a) broadly defines the outdoor aspects of the economy as agriculture, coastal recreational areas and natural landscapes – including the existence value of ecosystems, species and so forth. Consequently, indoor aspects of the economy covers what would remain out of the outdoor aspects.

Every CBA is subject to subjective uncertainty. In principle, this is not important because of the identical mathematical formulas in both subjective and objective cases. However, disguising their distinction leads identification of probability with exercises in calibration to sample frequencies from the data. Nevertheless, the climate change economics with its unknown unknowns demonstrates a notable application of subjective uncertainty theory (Weitzman, 2007a).

When we consider subjective uncertainty, we can still come up with a fair approximation of the central areas of the PDF. However, the challenges start when we start moving to the rear regions of the PDF, where we are in the unknown territory of subjective uncertainty. Here the probability estimates of the PDFs become dispersed to a great extent. This ambiguity then results in a fat left tail of the probability distribution of g. As a result, mitigation does not only move the center of the PDF of g to the right but also thins the left fat tail of the distribution.

To show the effect of fattened-tails Weitzman considers a simple example. In this example, uncertainties due to the unavailability of a sufficiently large data set makes us unsure about the mean and variance of g. He then shows that the reduced-form situation results in g that is distributed as Student-t with thickened tails. Using the Student-t distribution, the expected marginal utility of an additional sure unit of consumption is infinity while it is $e^{-\mu\eta+\frac{1}{2}\eta^2\sigma^2}$ in the pre-warming era.

Weitzman (2007a) claims that this general result has significant economic consequences and argues that rather than using unreasonably low rates of δ and η (implying the low rates in the Stern Review), it would be better to go directly with the genuine concern that there is a possibility that global warming will ultimately lead to very high temperatures and environmental disasters. It may well be the case that the option value of waiting for better information about destructive tail events is insignificant because early discovery is impossible, is too expensive, happens too slow or nothing effective can be done to reverse global warming. Therefore, immediate down payment on catastrophe insurance by drastically cutting CO_2 emissions is essential (Weitzman, 2007a).

3 Dismal Theorem

Actions taken today have consequences that are not easy to undo and that will be observed in the distant future, and we are unsure about a stochastic process with known objective frequency and objective-frequency probabilities. According to Weitzman (2009b), a much more disturbing issue regarding expected utility analysis are the unknowns; structural uncertainty coupled with an economic inability to evaluate the tragic losses from extreme variations in the temperature. So, what is the rational economic response the expected utility theory can offer under extreme catastrophes with little but highly unknown probabilities? Moreover, how the thin-tailed CBA of climate change needs to be treated?

Weitzman (2009b) contrasts with the conventional wisdom that ignores extreme temperature-change probabilities. The paper, relying on Bayesian statistics, argues that when the small chances of extreme events are more uncertain and obscure, the impact on present discounted expected utility for a risk-averse agent gets more severe and crucial.

Weitzman defines climate sensitivity as a key indicator of the temperature rise caused by increases in atmospheric CO_2. Let $\ln CO_2$ and S_1 be the increase CO_2 in the atmosphere and climate sensitivity, respectively. The latter converts the natural logarithm of CO_2 into T, the equilibrium temperature response, by

$$T \approx \frac{S_1}{\ln 2} \ln CO_2.$$

Rather than working with S_1 Weitzman works with S_2, a generalized climate-sensitivity that includes heat-induced releases of the considerable volume GHGs currently sequestered in arctic permafrost commonly as methane, CH_4, which is a potent GHG. Thus, density function of S_2 has a longer and fatter tail compared to the density function of S_1.

Weitzman (2009b) takes a two-period model and a utility function in CRRA form (see Eq. (3)). Normalizing the present consumption to one, $c_0 = 1$, the growth of consumption between two periods is

$$g \equiv \frac{c_1}{c_0} - 1,$$

which from $\ln(1+g) \approx g$ and $c_0 = 1$ can be written as

$$g = \ln c.$$

Here, g is a RV and can be interpreted implicitly as being some transfer function of the form $g = F(T)$ so that from $g = \ln c, c = e^{F(T)}$. As before, T denotes the increase in temperature since Industrial Revolution. Weitzman takes $F(T) = \kappa - \phi T$ with known positive constants κ and ϕ. The inference mechanism is such that past realizations of g are used in looking at the effects of temperature rise and consumption, which afterwards are incorporated into a Bayesian updated posterior PDF of g.

Suppose that the mean of consumption growth, g, is known, but its standard deviation is unknown. While the true value of the standard deviation is unknown, it is assumed that some i.i.d. observations are available to make an estimate of it. Pursuing his analysis further, Weitzman finds that the agent would sacrifice an infinite amount today to obtain a sure unit of consumption in the future. Since this outcome is unbounded, Weitzman (2009b) argues that some mathematical mechanism is required to constrain it. He places a positive lower bound on consumption so that consumption can never get below this strictly positive amount. The lower bound is not necessarily arbitrary because it can be related to a value such as the value of statistical life (VSL). He later finds that the higher is the VSL, the higher becomes amount the economy would be willing to sacrifice today to obtain a sure unit of consumption in the future.

This result is valid for any relative risk averse utility function satisfying a positive elasticity of the marginal utility. Weitzman (2009b) argues that it is difficult to remove the underlying economic problem that expected discount factors could become arbitrarily large owing to the statistical inferences about limiting tail behavior. The only right way to circumvent this potential result is to have a priori information. If, for example, a specific uncertainty affects one tiny portion of an individual's or society's overall portfolio of assets, exposure is confined to that specific part and fat tails would not be a big issue. However, climate change makes the previous argument hugely irrelevant.

In conclusion, CBA of deeply uncertain catastrophes is more concerned with learning how thick the tail can be, the overall expense of thinning the fat-tail and how much can it be thinned. As a result, the research can be directed at understanding better the uncertainty concerning the less plausible situations. This demands well-funded studies and experiments about its feasibility, detrimental environmental side effects and cost-effectiveness of geoengineering choices to make the tail thinner.

Recently after Weitzman (2009b), William Nordhaus came up with a critique (Nordhaus, 2009) where he analyzes the range of its application. As he argues, the conditions that are required for the theorem to hold are limited and do not apply to a broad range of scenarios. In particular, Nordhaus (2009) arrives at the conclusion that the Dismal Theorem holds if the distribution is very fat-tailed or if the utility function shows very high risk aversion. Hence, fat-tails by itself are not sufficient to cause infinite expected utility. Furthermore, finiteness depends on both the parameters of the utility function and parameters of the preference function. The exponential distribution that Weitzman puts as fat-tailed causes a finite expected utility or marginal utility in all cases. However, he agrees that Weitzman points to important issues concerning the choice of distributions when making decisions under uncertainty.

4 Additive Damages, Thick-Tailed Climate Dynamics and Uncertain Discounting

In Weitzman (2011), the author once more discusses the economic implications of large structural uncertainties that surround climate-change. He suggests that an additive form, in which welfare is the difference between the utility obtained from consumption and disutility (in quadratic form) accruing from temperature rise can also be used when assessing extreme climate damages.

One of the central questions in Weitzman (2011) is about the welfare evaluation of uncertain temperatures trajectories. He uses a simple energy model to evaluate how the PDF of temperature rise in time approaches its limiting PDF. Considering an additive damage function, welfare assessment mainly turns out to be the calculation of the discounted sum of the future utilities. When there is uncertainty in the discount rate which has a distribution with infinite small probability around zero, then the expected and discounted disutility due to temperature rise, and therefore, environmental damages, is infinite in the limit.

In the literature of climate change, the temperature damages can both be in an additive or multiplicative form. Such differences can cause significant differences in climate change policies. Weitzman (2011) argues that an additive form can be as reasonable as a multiplicative form. He assumes that the utility obtained from consumption is isoelastic with $\eta = 2$. The utility function in multiplicative form was already given by Eq. (10).

The damage function in additive form is given by Eq. (9). It is expected that the willingness to pay (WTP) to reduce temperature rise since the pre-warming period to zero will show differences. The question is what is the amount as fraction of current consumption (take $c_0 = 1$) that the representative agent would be willing to pay to reduce the temperature rise to zero, that is, $T(t) = 0$? Let this as amount be ω. As before, δ is the rate of pure time preference. Then, ω satisfies

$$u(1,0) - u(1-\omega,0) = e^{-\delta t} u(c_t, T_t).$$

If the consumption is supposed to grow at the rate g; and therefore, $c_t = e^{gt}$, the WTPs coming from the utility functions in multiplicative and additive forms, ω_m and ω_a, respectively, are

$$\omega_m = \frac{e^{[-(g+\delta)t]} \theta T_t^2}{1 + e^{[-(g+\delta)t]} \theta T_t^2} \text{ and } \omega_a = \frac{e^{[-\delta t]} \theta T_t^2}{1 + e^{[-\delta t]} \theta T_t^2}.$$

The additive formulation is free of the multiplicative term e^{-gt}. Numerical exercises show that while the WTP to avoid $T_{150} = 4°C$ under the multiplicative specification at time t = 0 is $\omega_m = 0.4$ %, it is $\omega_a = 7.5$ % for the additive specification. One lesson that one can have from this simple numerical exercise is that a seemingly unimportant difference between additive and multiplicative formulations can lead to significant differences. This is another example that structural uncertainty (coming from functional form of utility damages) can have on climate-change policy.

For one more time, Weitzman shows that the WTP to avoid climate change is unbounded and argues again that there are several plausible ways to avoid this inconvenient outcome. The disturbing infinite limit can be eliminated using ad hoc inequality constraints. However, this would not remove the underlying problem, because the outcome would then be an arbitrary large expected present discounted utility, whose exact value would depend on the obscure bounds, truncations, severely-dampened or cut-off prior PDFs, or some other formal mechanisms (Weitzman, 2011).

Weitzman concludes that the infinite amount of disutility should not discourage the climate change economists, nor avoid them from pursuing further studies and conduct new numerical simulations with policy implications. However, it is essential to note that CBAs in the climate change literature can relatively be more subject to subjective judgments about structural uncertainties.

Conclusion

Ideas in discounting and our understanding of climate change are evolving. Previously, the discussion over discounting was mainly ethical. Martin Weitzman has incorporated a new dimension to the economics of climate change literature by questioning the role of discounting under structural uncertainty: the resulting distribution of economic growth with a thickened left tail obscured the ethical debate and urged for immediate and strong action against global warming. Moreover, consumption inequality among generations has also become a second-degree issue (although this has not been scrutinized in the recent works of Weitzman).

Although Nordhaus acknowledges that Weitzman points to important issues concerning the choice of distributions when making decisions under uncertainty, he argues that the fat-tails by itself are not sufficient to cause infinite expected utility and, (in-)finiteness depends on both the parameters of the utility function and parameters of the preference function.

Another type of structural uncertainty comes from the utility functions. By employing utility function in additive and multiplicative forms, Weitzman has demonstrated significant differences between WTPs to prevent a certain amount of temperature rise. When the adverse of temperature rise also affects consumption growth (implying diminishing production possibilities frontier), damages health and decreases the amenity from the environment, Weitzman's result changes radically. The outcome is such that the difference between the WTPs gets smaller. This result, nevertheless, adds more uncertainty to the issue and provides another argument supporting structural uncertainty.

Other than the debate that has been going on between Martin Weitzman and William D. Nordhaus, the economics of climate change literature has seen other attempts concerning economic modeling and numerical techniques used (e.g., stochastic discount rates, IAMs incorporated with thickened tails of the PDF of economic growth and so on). Some prominent economists in this regard are Robert D. Pindyck and John Geanakoplos.

To conclude, the economics of climate change is evolving rapidly, and the recent studies point to the role of structural uncertainty in the analysis. It is clear that there is too much uncertainty surrounding climate change, but it is evident that signing very costly insurance contracts that aim at protecting ourselves from mega-catastrophes with tiny probabilities in the future is not realistic as well as applicable. Nevertheless, the enthusiasm to include structural uncertainty into the economic analysis in climate change enforces us to improve our knowledge in this issue (a point that is highly emphasized by Weitzman), and reduce the set of unknowns and work with more knowns.

References

Conceição, P. and Zhang, Y. (2010) "Discounting in the Context of Climate Change Economics: The Policy Implications of Uncertainty and Global Asymmetries", Environmental Economics and Policy Studies, 12(1–2): 31–57.

Dasgupta, P. (2008) "Discounting Climate Change", Journal of Risk and Uncertainty, 37: 141–169.

Douglas, M. and Wildavsky, A. (1982) "Risk and Culture (An Essay on the Selection of Technological and Environmental Dangers)", United States: University of California Press.

Farmer, J. D. and Geanakoplos, J. (2009) Hyperbolic Discounting is Rational: Valuing the Far Future with Uncertain Discount Rates, Cowles Foundation Discussion Papers No: 1719.

Hume, D. (1739) A Treatise on Human Nature: Being an Attempt to Introduce the Experimental Method of Reasoning into Moral Subjects, Vol. I [-III], 1739, John Noon.

IPCC (2014) "Climate Change 2014: Synthesis Report. Contribution of Working Groups I, II and III to the Fifth Assessment Report of the

Intergovernmental Panel on Climate Change [Core Writing Team, R.K. Pachauri and L.A. Meyer (eds.)]", Geneva, Switzerland.

Lucas Jr, R. E. (1978) "Asset Prices in an Exchange Economy", Econometrica: Journal of the Econometric Society, 46 (6): 1429–1445.

Nordhaus, W. D. (2007) "A Review of the Stern Review on the Economics of Climate Change", Journal of Economic Literature, 45(3): 686–702.

Nordhaus, W. D. (2009) "An Analysis of the Dismal Theorem", Cowles Foundation Discussion Papers No. 1686.

Nordhaus, W. D. (2016) "Projections and Uncertainties about Climate Change in an Era of Minimal Climate Policies", Technical Report, National Bureau of Economic Research No. w22933.

Pindyck, R. S. (2012) "Uncertain Outcomes and Climate Change Policy", Journal of Environmental Economics and Management, 63(3): 289–303.

Schelling, T. C. (1995) "Intergenerational Discounting", Energy Policy, 23: 395–401.

Stern, N. (2007) "The Economics of Climate Change: The Stern Review", Cambridge, United Kingdom: Cambridge University Press.

Sterner, T. and Persson, U. M. (2008) "An Even Sterner Review: Introducing Relative Prices into the Discounting Debate", Review of Environmental Economics and Policy, 2(1): 61–76.

Weitzman, M. L. (2007a) "A Review of the Stern Review on the Economics of Climate Change", Journal of Economic Literature, 45(3): 703–724.

Weitzman, M. L. (2007b) "Subjective Expectations and Asset-Return Puzzles", The American Economic Review, 97: 1102–1130.

Weitzman, M. L. (2009a) "The Extreme Uncertainty of Extreme Climate Change: An Overview and Some Implications", Working paper, Department of Economics, Harvard University.

Weitzman, M. L. (2009b) "On Modelling and Interpreting the Economics of Catastrophic Climate Change", The Review of Economics and Statistics, 91(1): 1–19.

Weitzman, M. L. (2009c) "Some Basic Economics of Extreme Climate Change", In Touffut, J.-P., Editor, Changing Climate, Changing Economy, Chapter 5, Massachusetts, United States: Edward Elgar Publishing.

Weitzman, M. L. (2011) "Additive Damages, Fat-Tailed Climate Dynamics, and Uncertain Discounting", In Libecap, G. D. and Steckel, R. H., Editors, The Economics of Climate Change: Adaptations Past and Present, 1: 23–46, Chicago, United States: University of Chicago Press.

Ayşe Durgun Kaygısız

The Correlation between Environmental Pollution and Economic Growth: Validity Analysis of the Environmental Kuznets Curve According to the Panel Data Method[1]

Introduction

Environmental issues are among globally important current issues. Emission of many hazardous gases such as carbon emission resulting from use of fossil fuels due to industrialization has been causing environmental pollution along with economic development lately. Furthermore, global warming and climate change problems draw attention to this field.

As environment was accepted as a financial variable, financial analyses including environmental pollution started to be conducted especially after the Second World War. Firstly, the Club of Rome stated in 1968 that uncontrolled use of resources due to population growth could damage the environment and pointed out that growth rates must be slowed down. Afterwards, the Kyoto Protocol that was signed in 1997 and entered into force in 2005 aimed to reduce greenhouse gas emissions of countries. In addition, the correlation between environment and growth is explained with the Environmental Kuznets Curve (EKC) in the economic theory.

The main objective of this study is to test the validity of the EKC hypothesis between 2003 and 2016 in upper-middle income countries located in the European-Central Asian territory[2]. Also the study aims to analyze impacts of electricity consumption and population density on EKC. We think that the study will make a significant contribution to the literature as the country group, variables and dataset are different from those in other studies. The first part of the study explains the EKC hypothesis while the second part mentions the domestic and foreign studies conducted in this field. The final part of the study discusses the dataset and econometric method and evaluates the findings.

---

1 This study is presented as a summary in International Conference on Emprical Economics and Social Sciences (ICEESS'18)
2 Albania, Azerbaijan, Belarus, Bosnia and Herzegovina, Bulgaria, Croatia, Kazakhstan, Macedonia, Montenegro, Romania, Russia, Serbia, Turkmenistan and Turkey.

1 Theoretical Approach: Environmental Kuznets Curve

Kuznets (1955) states that income inequality increases in early years of economic development of countries, but this injustice gradually decreases after a certain turning point provided that the economic development continues. This correlation between per capita income and income inequality is known as the Kuznets Curve and drawn as inverted U. Grossman and Krueger (1991–1995) replaced the axis showing income inequality with environmental pollution and reinterpreted the Kuznets Curve as the correlation between growth and environmental pollution. This curve is called the Environmental Kuznets Curve in the economy literature. The EKC hypothesis suggests that environmental pollution will increase in early years of economic growth, but pollution will decrease in the following years of the growth. This is explained as follows: Mass production and excessive increase in consumption due to industrialization increase the need for energy. This need increases external dependence of countries as well as use of fossil-based energy sources instead of efficient energy sources. Use of fossil-based energy sources increases the CO_2 greenhouse gas emission in the atmosphere. The production increase results in growth but growth increases CO_2 emission causing environmental pollution. Output amount is attached importance in early years of economic growth and therefore environmental problems are ignored at the cost of growth. Still, per capita income increasing in parallel with growth causes people to become conscious and environmental awareness to increase. As a result, it is predicted that pollution in developing countries will be temporary and environmental pollution will decrease at high growth levels. In this context, two main correlations between economic growth and environmental pollution are mentioned (Karaca, 2012: 140). The first one is the income elasticity between income and clean environment. Accordingly, life standards of individuals increase after a certain income level and they want to live in a cleaner environment (Panayotou, 2003: 46). Beckerman (1992: 91) confirms this suggestion by saying that the most certain way to improve the environment is to get rich. Furthermore, other studies on this matter have found out that high personal income and life standards increase the value attached to the environment (Pezzey, 1989; Selden and Song, 1994; Baldwin, 1995). The second factor is that laws related to environmental planning are stricter and desired legal regulations can be made more easily in rich countries. People with high income in such countries are more sensitive to the environment and they support laws aimed at increasing measures to protect the environment.

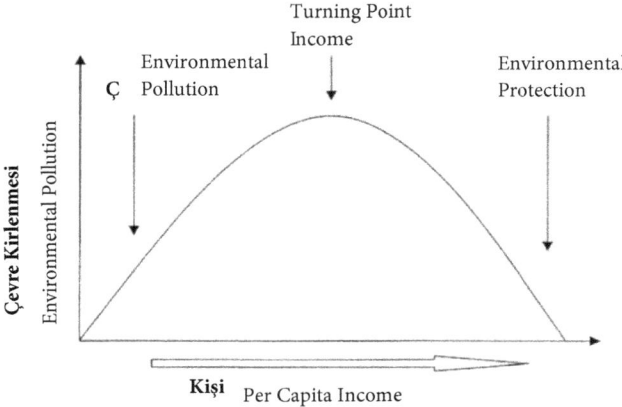

Fig. 1: Environmental Kuznets Curve. Source: Yandle et al., 2004: 3

As you can see in Fig. 1, the Environmental Kuznets Curve is drawn as an inverted U. This is theoretically explained as scale effect, structural effect and technological effect (Ang, 2007: 4773). Scale effect is associated with growth-based production increase and formation of environmentally hazardous waste in parallel with it. Therefore, the increasing part of EKC is explained with scale economy. The decreasing part of EKC is explained with structural effect and technological effect. Economic structure changes in time. Although industrial pollution occurs during transition from agricultural society to industrial society, environmental pollution decreases during transition from industrial society to service and information society. Also usage of environmentally friendly and clean technologies in developing societies helps to increase the environmental quality. All these cause EKC to have a decreasing trend.

CO_2 is one of the most important indicators of environmental pollution accepted in the economics literature. CO_2 has the largest share with 58.8 % among greenhouse gases causing pollution (Altıntaş, 2013: 264). Therefore, CO_2 emission is used to represent environmental pollution in most EKC studies. Furthermore, different variables have been added to basic variables in the literature. As energy consumption is a factor affecting environmental pollution (Narayan and Narayan, 2010: 661; Richmond and Kaufmann, 2006; Apergis and Payne, 2009a, 2010; Pao and Tsai, 2010; Soytaş et al., 2007; Öztürk and Uddin, 2012, 2016; Altıntaş, 2013), this variable is also added to the hypothesis. As increase in energy consumption will increase economic activities, it indicates a growing economy (Apergis and Payne, 2009a, 3282–3285). Increase in energy consumption will cause an increase in carbon emission in addition to increase in growth and production. Moreover, the energy need will increase as growth increases and this will increase CO_2 emission. In the light of all these factors, a positive correlation is expected between energy consumption and CO_2. Population density data is also added to the EKC hypothesis to measure the impact of population increase on environmental pollution (Panayotou, 1993; Franklin and Ruth, 2012; Zeren and Arı, 2011; Karakaş, 2016). The impact of population on environment is interpreted in two ways in the literature. Firstly, population increase will accelerate excessive use of natural resources and energy consumption, thus increasing the damage to the environment. Those with an opposing view argue that environmental concerns will increase along with increased population and this will reduce environmental destruction by setting the environmentalist movement in motion (Şahinöz and Fotourehchi, 2013: 206). As a result, it is not right to say anything for sure about the correlation between population and environment. Nevertheless, the dominant view holds that the negative impact of population increase is much higher than the positive impact.

2 Literature Review

A lot of studies have been conducted in both domestic and foreign literature on the EKC theory that represents sustainable development by modeling the correlation between economy and environment. Different results have been found in these studies. That is because they have been conducted with different models, variables, country groups, analyses and periods. These studies are summarized below.

Şahinöz and Fotourehchi (2013) tested the validity of EKC in OECD countries according to reduced and decomposed models. An N type correlation was detected between CO_2 and GDP in that study covering the period from 1994 to 2010. They also state that environmental policies shift EKC downwards and decrease the slope. Another study was conducted by Aytun et al. (2017) in this field. They conducted the study on 10 developing countries for the period from 1980 to 2010 and used the Pedroni Co-integration and Fully Modified Ordinary Least Squares (FMOLS) estimation methods. The results confirm the EKC hypothesis as well as revealing the negative effect of electricity usage on environmental pollution. Another study analyzing the correlation between growth and environmental pollution was conducted by Karakaş (2016). The study included 61 countries in different income groups and used the panel data method for the period between 1990 and 2013.

A strong correlation was detected between population income level and CO_2. Another study conducted on OECD countries covers the period between 2000 and 2012. The study conducted by Gülmez (2015) investigated the correlation between growth and air pollution with co-integration, (Fully Modified Ordinary Least Squares, Dynamic Ordinary Least Squares (FMOLS, DOLS) and panel causality tests. He found out that these two variables were co-integrated in the long term and the air pollution coefficient was 3 % in average. Öztürk and Yıldırım (2015) analyzed the period between 1967 and 2010 in their study on MINT (Mexico, Indonesia, Nigeria and Turkey) countries. They conducted a long-term panel causality analysis and found out that EKC was supported only for Nigeria. Erataş and Uysal (2014) tested the correlation between environmental pollution and income level in BRICST (Brazil, Russia, India, China, South Africa and Turkey) countries. They used the panel data method and confirmed the EKC hypothesis. Another study conducted by Gündüz (2014) included OECD countries in the period between 1960 and 2008. Panel co-integration analysis was conducted and a long-term correlation was detected between environmental pollution and growth in that study. Another study tested the EKC hypothesis using the data about growth, CO_2 and population density in a total of 30 developed and developing countries between 1992 and 2009. Sarısoy and Yıldız (2013) used the panel data analysis method in that study. They detected an N-type correlation between the variables. Güriş and Tuna (2011) used the parametric panel data model for 88 countries and confirmed the EKC hypothesis in those countries for the period between 1971 and 2008. Arı and Zeren (2011) examined Mediterranean countries and analyzed the correlation between CO_2 and growth according to the panel data method. They found out that an N-type correlation existed between CO_2 and per capita growth, and population and energy consumption had a positive impact on CO_2. Turkey and the European Union were compared in terms of the EKC hypothesis in a study covering the period from 1968 to 2003. Kotil et al. (2009) used the Grey model in that study. According to the results, the emission amount increased along with the income increase in Turkey while the increase in income reduced the emission amount in the European Union.

Pao and Tsai (2010) analyzed the dynamic panel causality correlations among the output amount, pollutive emissions and energy usage in BRIC (Brazil, Russia, India, China) countries (except the period between 1990 and 2005 for Russia) in the period from 1971 to 2005. They found a positive and significant correlation between the variables in the long term. In other words, they found out that energy usage had a strong bidirectional correlation with both pollutive emissions and growth. Franklin and Ruth (2012) state that according to long term data covering a 200-year period, EKC was valid at first but it had a backfiring effect and CO_2 emission increased along with growth afterwards. They state that the EKC hypothesis was confirmed only in short term analyses during the period from 1900 to 2000. They also included data about old and young population in their study. They found out that CO_2 had a negative correlation with old population while it had a positive correlation with young population. Narayan and Narayan (2010) analyzed the period from 1980 to 2004 for 43 developing countries. They used the panel co-integration and panel long term estimation techniques and confirmed the validity of the EKC hypothesis for 35 % of the countries. Lean and Smyth (2010) analyzed the causality relationship among CO_2 emission, economic growth and energy consumption for the period from 1980 to 2006 in five ASEAN countries using the panel vector error correction model. They determined a statistically significant correlation between electricity consumption and emissions and a non-linear correlation between emissions and growth in line with the EKC hypothesis in the long term. Selden and Song (1994) tested EKC in 22 OECD and 8 developed countries for the period between 1979 and 1987. They obtained results confirming the hypothesis. Shafik and Bandyopadhyay (1992) analyzed 10 different variables as indicators of pollution for 135 countries and found out that pollution levels increased in all variables except water pollution along with the increase in income. Panayotou (1993) confirmed the EKC hypothesis in a study taking air pollution and desertification rate into account for developed and developing countries. In addition, he found out that population density increased the level of desertification. Grossman and Krueger (1995) investigated the correlation between air pollution and growth for 42 countries and found out that growth impaired the environmental quality but the environmental quality increased in following periods. Richmond and Kaufmann (2006) tested the EKC hypothesis by including energy usage in different country groups. The hypothesis was confirmed in OECD countries whereas no change was detected among variables in countries outside OECD. Energy usage was included in the EKC hypothesis in another study by Apergis and Payne (2009b). They used the panel error correction model and conducted the causality test on six central American countries for the period from 1971 to 2004. They found a positive correlation among variables in the long term and confirmed the EKC hypothesis. They determined a unidirectional causality between energy consumption and growth in the short term. Another study by Apergis and Payne (2010) included 11 members of the commonwealth of independent states for the period from 1992 to 2004. They analyzed the causality relationship among real output, CO_2 emissions and energy consumption using the panel vector error correction model. They found out that energy usage had a positive impact on CO_2 emission in the long term and confirmed the EKC hypothesis. They determined a unidirectional causality between energy consumption and real output in the short term. Zhang and Cheng (2009) investigated the causality relationship among economic growth, carbon emission and electricity consumption for the Chinese economy. The study covered the period from 1960 to 2007 and they determined a unidirectional causality relationship between energy consumption and carbon emission in the long term. Ang (2007) investigated the correlation among CO_2 emission, energy consumption and total production for France between 1960 and 2000. According to the results of the causality test, growth had a positive correlation with energy consumption and CO_2 emission in the long term.

3 Dataset, Econometric Method and Evaluation of the Findings

This part introduces the variables and analysis method used in the study. Also the findings are analyzed.

3.1 Dataset and Econometric Method

A balanced panel data analysis was conducted in the study. Although panel data analysis encompasses qualities specific to both time-series and cross section data analyses, it can also eliminate disadvantages of such analyses (Tarı, 2011: 475). The fact that panel data possesses both time and cross section qualities increases information usage and the degree of freedom. As the number of observations increase, more variables are added to the measured correlation, thus eliminating multiple linear connection problems (Hsiao, 2003: 7).

Both the fixed effects model and the random effects model were used in the study. The best choice between fixed effects and random effects for panel data analysis is to investigate whether the parameters of both models are in correlation with individual effects or unobservable effects (Tarı, 2011: 494). The model was chosen with Hausman test in the study.

The correlation between environment and income is tested with three basic models in the literature (Shafik and Bandyopadhyay, 1992). These models are as follows:

$$E_{i,t} = \beta_1 + \beta_2 Y + \beta_3 Z + \varepsilon_{i,t} \qquad (1)$$

$$E_{i,t} = \beta_1 + \beta_2 Y + \beta_3 Y^2 + \beta_4 Z + \varepsilon_{i,t} \qquad (2)$$

$$E_{i,t} = \beta_1 + \beta_2 Y + \beta_3 Y^2 + \beta_4 Y^3 + \beta_5 Z + \varepsilon_{i,t} t \qquad (3)$$

E, Y and Z variables in these models represent factors other than income affecting environmental pollution, income per capita and environmental quality respectively while ε_t represents the error term. Models 1, 2 and 3 estimate the linear, quadratic and cubic correlation between environment and income respectively. In the quadratic model, an inverted U correlation is obtained between income and environmental pollution if β_2 is positive and is β_3 negative. For this correlation to exist in the cubic model, β_2 must be positive, β_3 must be negative and β_4 must be zero.

The quadratic model (Selden and Song, 1994; Şahbaz et al. 2012; Aytun et al., 2017; Atıcı and Kurt, 2007) or cubic model (Dinda, 2004; Arı and Zeren, 2011; Karaca, 2012) has been used in most empirical studies. A quadratic model was used in this study. The model was defined in accordance with the models used in previous studies and specified as follows. Furthermore, the turning point was calculated with the Y=β1/2β2 formula (Kılıç and Akalın, 2016: 51).

$$CO_{2it} = \beta_0 + \beta_1 Y_{it} + \beta_2 Y_{it}^2{}_{it} + \beta_3 EN_{it} + \beta_4 NFS_{it} + \varepsilon_{i,t} \qquad (4)$$

This study aims to test the validity of the EKC hypothesis in upper-middle income countries in Central Asia and Europe between 2003 and 2016. The dependent variable of the study is CO_2 emission per capita while the independent variable is the growth rate in the GDP per capita. Also per capita energy consumption and population growth rate were added to the model as control variables. Logarithms were taken for the data without a ratio. All observations were obtained from the World Development Indicators database of the World Bank. The study was limited to the period between 2003 and 2016 as 2017 data had not been announced yet.

3.2 Evaluation of Findings

Observation number, mean, minimum and maximum value data of the model etc. are given in the statistics summary table.

Correlation is important for reliability of the model estimation (Güngördü, 2002: 64). Therefore, the correlation among the variables is shown in Tab. 3. The correlation results show that the highest correlation value is 78 %. It was determined that nothing could cause any problems in choosing the variables for the model as the correlation among the variables was not high.

The modified Wald test was conducted to test the heteroscedastic problem of the model. The test results are shown in Tab. 4. As you can understand from the table, the Ho hypothesis was rejected. The robust method was used to correct the heteroscedastic problem in the model and the problem was solved.

Tab. 1: Used Variables and Expected Impact. Source: Created by Authors

Name of the Variable	Explanation	Database	Expected Impact
CO_2	Per capita carbon dioxide in kg	World data bank	-
Y	Growth rate in GDP per capita	World data bank	Positive
Y^2	Square of the growth rate in GDP per capita	Calculated by us	Negative
EN	Per capita petroleum equivalent of energy consumption	World data bank	Positive
NFS	Population growth rate	World data bank	Positive

Tab. 2: Statistics Summary Table. Source: Created by Author

Variables	Obs	Mean	Std. Dev.	Min	Max
CO_2	196	1.013967	0.6517607	0.2939675	3.839212
Y	196	4.356361	4.851515	-7.84864	33.03049
Y^2	196	42.394	100.3308	0.0073258	1091.013
EN	196	7.644776	0.5119678	6.474151	8.55005
NFS	196	0.172438	0.8432996	-1.666383	2.636253

Tab. 3: Correlation Results. Source: Created by Authors

	CO_2	Y	Y^2	EN	NFS
CO_2	1.000				
Y	0.2840	1.000			
Y^2	0.1543	0.7644	1.000		
EN	0.6089	0.0584	0.0172	1.000	
NFS	0.2924	0.1402	0.1598	0.2271	1.000

Tab. 4: Modified Walt Test Results. Source: Created by Authors

chi2 (14) = 120.40
Prob>chi2 = 0.0000

Tab. 5: Wooldridge Test. Source: Created by Authors

$F(1, 13) = 1.532$	
$Prob > F = 0.2377$	

Tab. 6: Panel Data Estimation Table. Source: Created by Authors

Variables	CO_2	CO_2	CO_2	CO_2
Y	0.0394***	0.0661***	0.0539***	**0.0422***
	(0.0113)	(0.0193)	(0.0167)	**(0.0115)**
Y^2		-0.00136**	-0.00136**	**-0.000908***
		(0.000608)	(0.000521)	**(0.000292)**
NFS			0.211***	**0.114***
			(0.0328)	**(0.0375)**
EN				**0.730***
				(0.0298)
Constant	0.842***	0.783***	0.800***	-4.732***
	(0.0493)	(0.0614)	(0.0531)	(0.226)
Obs	196	196	196	196
R^2	0.054	0.069	0.142	0.470

Robust standard errors in parentheses.
***p<0.01, **p<0.05, *p<0.1

The Wooldridge test was conducted to test the autocorrelation problem of the model. The test results are shown in Tab. 5. The Ho hypothesis was accepted. No autocorrelation problem was observed in the model.

Fixed and ransom effects were analyzed in the model and a selection was made between the tests by conducting the Hausman test. The panel data estimation findings according to the analysis results are shown in Tab. 6. Also analysis results of regression, fixed and random effects are given in the Appendix.

As you can see in Tab. 6, all variables are significant by 1 %. Also the estimation results in expected effects are consistent.

The shape of the EKC function is determined by signs of the coefficients belonging to Y and Y^2. EKC can be U or inverted U depending on the coefficients. The sign of coefficient Y is positive while the sign of Y^2 is negative. In this case, the hypothesis that EKC is in the shape of inverted U was supported in this study. In other words, CO_2 emission increased along with the increase in the growth rate of GDP per capita. The turning point of EKC was determined as 23.24

growth rate. 1 % increase in Y increases CO_2 by 0.0422 %. But the CO_2 emission amount decreases when GDP per capita reaches the 23.24 growth rate that is the turning point. 1 % increase in GDP per capita after the determined turning point will decrease CO_2 emission by 0.001 %.

The energy consumption coefficient is also positive and statistically significant. In this context, energy consumption increases CO_2 emission as expected in theory. Similarly, the population growth coefficient is positive and statistically significant. CO_2 emission increases along with the increase in population.

Conclusion

The world economy has been growing rapidly lately. The fast growth can cause some environmental problems. Pollution resulting from using more natural resources for higher production and fossil fuels for energy is the main reason for environmental problems. Environmental problems have accumulated in time and started to pose a threat for human life. This reveals the importance of sustainable development. The correlation between environment and economy became one of the current topics in the academic environment especially after 1990 and a lot of studies have been conducted since then. In this sense, the EKC hypothesis attained a place in the literature by explaining the correlation between environment and economy in academic terms.

The correlation between economic growth and CO_2 emission in upper-middle income countries located in Europe and Central Asia for the period between 1993 and 2016 was tested in this study using the panel data analysis method. Electricity consumption and population growth rate were added to the model as control variables. The EKC hypothesis was determined to have an inverted U correlation and the validity of the hypothesis was accepted. An increase of 1 % in Y increases CO_2 by 0.0422 %. On the other hand, when GDP per capita reaches 32.24 growth rate that is the turning point, 1 % increase in GDP per capita will decrease CO_2 emission by 0.001 %. It was found out that the energy usage and population increase variables used in the study had a positive correlation with CO_2 emission. As the energy usage increases, CO_2 emission will increase as well. In other words, increase in energy consumption will cause an increase in carbon emission in addition to increase in growth and production. Moreover, the energy need will increase as growth increases and this will increase CO_2 emission. Increase in the number of people in countries also increases CO_2 emission. Population increase will accelerate excessive use of natural resources and energy consumption, thus increasing the damage to the environment.

Economic growth initially causes environmental pollution in developing countries that aim to achieve economic growth at a high level. However, the countries will develop, the demand for a high-quality environment will increase, an environmental awareness will develop and the environmental pollution rate will decrease as the growth increases. Economic growth at a high level will be beneficial to the environment. But there are other issues that must be paid attention. That is because environmental pollution cannot be explained only with personal income. Therefore, energy consumption and population amount were added to the study. In this context, both renewable energy resources and raising awareness in this matter are important in the economic context in which the environment becomes more important day by day. For this reason, countries must revise their energy and population policies and make legal arrangements that reduce environmental pollution and are in harmony with the sustainable growth objective. For example, costs of renewable energy resources must be reduced and bureaucratic obstacles regarding this matter must be minimized as in many European Union countries. Moreover, quota requirements may be imposed on companies generating electricity such as generating a certain amount of energy from renewable resources.

Appendix

Tab. 7: Fixed Effects. Source: Created by Authors

Variables	CO_2	CO_2	CO_2	CO_2
KBMG	0.0394***	0.0661***	0.0539***	0.0422***
	(0.0123)	(0.0198)	(0.0194)	(0.0153)
KBMG		-0.00136*	-0.00136*	-0.000908
		(0.000793)	(0.000763)	(0.000603)
NFS			0.211***	0.114***
			(0.0539)	(0.0435)
ENRJ				0.730***
				(0.0696)
Constant	0.842***	0.783***	0.800***	-4.732***
	(0.0704)	(0.0780)	(0.0752)	(0.531)
obs	196	196	196	196
R^2	0.054	0.069	0.142	0.470

Standard errors in parentheses. ***$p<0.01$, **$p<0.05$, *$p<0.1$

Tab. 8: Random Effects. Source: Created by Authors

Variables	CO_2	CO_2	CO_2	CO_2
KBMG	0.0381***	0.0537***	0.0521***	0.0446***
	(0.00925)	(0.0143)	(0.0138)	(0.0112)
KBMG		-0.000981	-0.00120*	-0.000856
		(0.000692)	(0.000669)	(0.000541)
NFS			0.207***	0.108***
			(0.0518)	(0.0429)
ENRJ				0.713***
				(0.0699)
Constant	0.848***	0.822***	0.802***	-4.614***
	(0.0602)	(0.0628)	(0.0607)	(0.533)
obs	196	196	196	196

Standard errors in parentheses. ***p<0.01, **p<0.05, *p<0.1

References

Altıntaş, H. (2013) "Türkiye'de Birincil Enerji Tüketimi, Karbondioksit Emisyonu ve Ekonomik Büyüme İlişkisi: Eşbütünleşme Ve Nedensellik Analizi", Eskişehir Osmangazi Üniversitesi, İİBF Dergisi, 8(1): 263–294.

Ang, J. B. (2007) "CO_2 Emissions, Energy Consumption, and Output in France", Energy Policy, 35(10): 4772–4778.

Apergis, N. and Payne, J.E. (2009a) "CO_2 Emission, Energy Usage and Output Central Amerika", Energy Policy, 37: 3282–3286.

Apergis, N. and Payne, J.E. (2009b) "Energy Consumption and Economic Growth in Central Amerika: Evidence from a Panel Cointegration and Error Correction Model", Energy Econ, 31: 211–216.

Apergis, N. and Payne, J.E. (2010) "The Emission Energy Consumption and Growth Nexus: Evidence from the Common Wealth of Independent State", Energy Policy, 38: 650–655.

Arı, A. ve Zeren, F. (2011) "CO_2 Emisyonu ve Ekonomik Büyüme: Panel Veri Analizi", Yönetim ve Ekonomi, 18(2): 37–47.

Atıcı C. ve Kurt, F. (2007) "Türkiye'nin Dış Ticaret ve Çevre Kirliliği: Çevresel Kuznets Eğrisi Yaklaşımı", Tarım Ekonomisi Dergisi, 13(2): 61–69.

Aytun, C., Akın, C.S. ve Algan, N. (2017) "Gelişen Ülkelerde Çevresel Bozulma, Gelir ve Enerji Tüketimi İlişkisi", Ömer Halis Demir Üniversitesi, İİBF Dergisi, 10(1): 1–11.

Baldwin, R. (1995) "Does Sustainability Require Growth?", I. Goldin ve L.A. Winters (ed), The Economics of Sustainable Development içinde. Cambridge: Cambridge University Press.

Beckerman, W. (1992) "Economic Growth and the Environment: Whose Growth? Whose Environment", World Development, 20(4): 481-492.

Dinda, S. (2004) "Environmental Kuznets Curve Hypothesis: A Survey", Ecological Economics, 49(4): 432-455.

Erataş, F. ve Uysal, D. (2014) "Çevresel Kuznets Eğrisi Yaklaşımının "BRICT" Ülkeleri Kapsamında Değerlendirilmesi", İktisat Fakültesi Mecmuası, 64: 1-25.

Franklin, R. S. and Ruth, M. (2012) "Growing up and Cleaning up: The Environmental Kuznets Curve Redux", Applied Geography, 32(1): 29-39.

Grossman, G.M. and Krueger, A. (1991) "Environment Impact of a North American Free Trade Agreement", NBER Research Working Paper, No: 3194, Cambridge.

Grossman, G.M. and Krueger, A. (1995) "Economic Growth and Environment", Quarterly Journal of Economics, 110(2): 353-377.

Gülmez, A. (2015) "OECD Ülkelerinde Ekonomik Büyüme ve Hava Kirliliği İlişkisi: Panel Veri Analizi", Kastamonu Üniversitesi İİBF Dergisi, 9: 18-30.

Gündüz, H.İ. (2014) "Çevre Kirliliği ile Gelir Arasındaki İlişkinin İncelenmesi: Panel Eşbütünleşme Analizi ve Hata Düzeltme Modeli", Marmara Üniversitesi, İİB Dergisi, 36(1): 409-423.

Güngördü, E. (2002) "Coğrafya'da İstatistik Methodları, 1. Baskı, Nobel Yayın Dağıtım", Ankara.

Güriş, S. ve Tuna, E. (2011) "Çevresel Kuznets Eğrisinin Geçerliliğinin Panel Veri Modelleriyle Analizi", Trakya Üniversitesi Sosyal Bilimler Dergisi, 13(2): 173-190.

Hsiao, C. (2003) "Analysis of Panel Data", 2nd Edition, Cambridge University Press, Published by The Press Syndicate of the University of Cambridge.

Karaca, Ç. (2012) "Ekonomik Kalkınma Çevre Kirliliği İlişkisi: Gelişmekte Olan Ülkeler Üzerine Ampirik Bir Analiz", Ç.Ü. Sosyal Bilimler Enstitüsü Dergisi, 21(3): 139-156.

Karakaş, A. (2016) "Yaklaşan Tehlikenin Farkına Varmak: İktisadi Büyüme, Nüfus ve Çevre Kirliliği İlişkisi", Selçuk Üniversitesi Sosyal Bilimler Meslek Yüksekokulu Dergisi, 41. Yıl Özel Sayı, 19: 57-73.

Kılıç, R. ve Akalın, G. (2016) "Türkiye'de Çevre ve Ekonomik Büyüme Arasındaki İlişki: ARDL Sınır Testi Yaklaşımı", Anadolu Üniversitesi Sosyal Bilimler Dergisi, 16(2): 49-60.

Kotil, E., Eryiğit, M. ve Konur, F. (2009) "Türkiye ve Avrupa Birliğinde CO_2 Emisyonu ve Gelir İlişkisi", Ekonomik Yaklaşım, 20(73): 55-67.

Kuznets, S. (1955) "Economic Growth and Income Inequality", American Economic Review, 45(1): 1-28.

Lean H.H. and Smyth, R. (2010) "CO_2 Emissions Electricity Consumption and Output in ASEAN", Applied Energy, 87(6): 1858-1864.

Narayan, P.K and Narayan, S. (2010) "Carbon Dioxide Emissions and Economics Growth: Panel Data Evidence from Developing Countries", Energy Policy, 38: 661-666.

Öztürk, İ. ve Salah Uddin G. (2012) "Causality among Carbon Emissions, Energy Consumption and Growth in India", Ekonomska İstrazivanja, 25(3): 752-775.

Öztürk, Z. ve Yıldırım, E. (2015) "Environmental Kuznets Curve in the Mint Countries: Evidence of Long Run Panel Causality Test", Ekonomik ve Sosyal Araştırmalar Dergisi, 11 (1): 175-183.

Panayotou T. (1993) "Empirical Test and Policy Analysis of Environmental Degradation at Different Stages of Economic Development", Working Paper WP238 Tecnology and Employment Programme, Geneva: İnternational Labour Organization, No: 292778.

Panayotou, T. (2003) "Economic Growth and the Environment", Economic Survey of Europe, No: 2

Pao, H.T. and Tsai, C.H. (2010) "CO_2 Emission, Energy Consumption and Economic Growth", Energy Policy, 38(12): 7850-7860.

Pezzey, J.C.V. (1989) "Economic Analysis of Sustainable Growth and Sustainable Development", Environment Department Working Paper 15, World Bank.

Richmond, A. K. and Kaufmann, R.K. (2006) "Is there a Turning Point ın the Relationship between Income and Energy Use and Carbon Emissions?", Ecological Economics, 56(2): 176-189.

Sarısoy, S. ve Yıldız, F. (2013) "Karbondioksit Emisyonu ve Ekonomik Büyüme İlişkisi: Gelişmiş ve Gelişmekte Olan Ülkeler İçin Panel Veri Analizi", Namık Kemal Üniversitesi Sosyal Bilimler Enstitüsü Sosyal Bilimler Metinleri Dergisi, 2: 1-22.

Selden T. ve Song, D. (1994) "Environment Quality and Development: Is there a Kuznets Curve for Air Pollution Emissions?", Journal of Environmental Economics and Management, 27(2): 147-162.

Shafik, N. and Bandyopadhyay, S. (1992) "Economic Growth and Environmental Quality: Time Series and Cross Section Evidence". World Bank Working Paper, WPS904, Washington D.C.

Shahbaz, M., Lean, H.H. and Shabbir, M.S. (2012) "Environmental Kuznets Curve Hypothesis in Pakistan: Cointegration and Granger Causality", Renewable and Sustainable Energy Review, 16(5): 2942-2953.

Şahinöz, A. ve Fotourehchi, Z. (2013) "Çevresel Kuznets Eğrisi: İndirgenmiş ve Ayrıştırılmış Modellerle Ampirik Bir Analiz", Hacettepe Üniversitesi İİBF Dergisi, 31(1): 199-224.

Tarı, R. (2011), Ekonometri, Umuttepe Yayınları: Kocaeli.

Yandle, B. and Bhattarai, Vijayaraghavan, M. (2004) "Environmental Kuznets Curve: A Review of Findings Methods and Policy İmplication", Research Study (2): 1–16.

Zhang, X.P. and Cheng X.M. (2009) "Energy Consumption, Carbon Emissions, and Economic Growth ın China", Ecological Economics, 68(10): 2706–2712.

İlyas Okumuş and Abdulmecit Yıldırım

Investigating the Validity of the EKC Hypothesis in Eurasian Countries: The Role of Financial Development

Introduction

Human beings have gone through various ordeals such as war, famine, rapid population growth and urbanization etc. in different territories throughout history. However, environmental problems, especially global warming and climate change that came into being in the 1970s affected all countries regardless of the development level. The reason for serious environmental problems such as global warming and climate change is the recent high increase in the amount of greenhouse gases. The amount and composition of greenhouse gases that have a natural balance in the structure of the atmosphere and determine how much of beams that come from the sun and reflect from the ground will go through the atmosphere started to change as of the Industrial Revolution. Mass production that started with the Industrial Revolution increased the population drastically and the rapid growth of population brought about a rapid increase in international commerce and the demand for energy which is the main input of urbanization, consumption and production. The energy need resulting from these developments was satisfied with fossil fuels such as coal, petroleum and natural gas. A significant amount of CO_2 emission mixes with the atmosphere as a result of fossil fuel consumption. This causes the CO_2 gas, which has the largest share among greenhouse gases, to gradually increase and disrupts the greenhouse gas composition that has a natural balance. Disruption of this balance limits the permeability of the atmosphere and affects global warming negatively.

Although the main reason for global warming and climate change is the increase in CO_2 emission, it is not the only cause of these problems. The rapid urbanization process together with the rapid industrialization has caused deterioration of natural resources and decreased the environmental quality. As people migrate from the country to cities, cities exceed their natural borders, causing the disruption of natural resources such as forests, farmlands and wetlands etc. This urbanization process causes the destruction of many life-sustaining natural areas. Furthermore, needs like food, energy and water emerging in such areas place an extra burden on the environment. In addition, an irreversible destruction happens in the ecosystem as waste produced in cities is thrown into nature unconsciously. All these problems increase the global warming problem.

The liberalization wave that has taken place in the world trade recently is also considered to be a significant factor increasing the pressure on the ecosystem. According to neoliberal finance, free trade is essential for increasing the welfare of countries and ensuring development. Due to trade liberalization, countries overproduce to increase their trade volumes. Overproduction causes natural resources and energy to be used excessively. In addition, the destruction of natural resources increases and environmental problems occur.

People have not remained unresponsive to these recent undesirable developments. Serious environmental problems have started to be discussed in many platforms and created new theoretical and political perspectives in many disciplines. Many steps are being taken and many organizations have been founded in parallel with these theoretical and political developments. This reactive model that has an important place in many interdisciplinary fields and finds conventional development models insufficient is called sustainable development. In general, sustainable development is a model with economic, social and environmental dimensions that address economic growth and ecological balance at the same time, attaches importance to environmental quality, advocates the efficient use of natural resources and suggests that current needs must be met by taking into account the needs of future generations.

As environmental problems became global, countries held meetings under the auspices of United Nations (UN) to take preventive measures and put the sustainable development model in effect. For example, an important step was taken in 1997 for reducing the amount of greenhouse gases to a certain level all over the world with the protocol prepared after the Kyoto meeting that was the third Conference of the Parties. Some restrictions were imposed on the emission of greenhouse gases with this protocol that entered into force in February 2005. It was pointed out in that agreement that regulations must be made on matters such as ensuring efficiency of energy consumption, using renewable energy sources instead of fossil fuels, using technologies requiring less energy in industry, transferring such technologies to all countries, imposing additional taxes on all countries consuming much energy and clearing the way for solar energy etc. Similarly, many agreements have been made for steps that must be taken for the sustainable development model. In order to achieve these objectives, firstly economic development factors

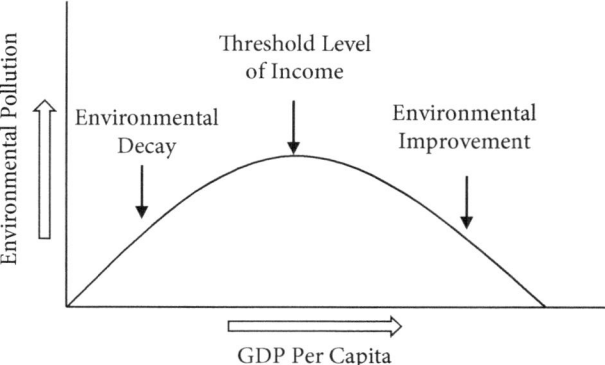

Fig. 1: Environmental Kuznets Curve. Source: Yandle, Bhattarai, and Vijayaraghavan, 2004

increasing environmental degradation must be determined properly and their impacts must be calculated.

The remainder of the study is organized as follows. Section 1 and Section 2 introduce the EKC hypothesis and literature review, respectively. Section 3 introduces dataset and their sources, and the econometric model with some descriptive statistics. Section 4 deals with the empirical findings. The final section concludes.

1 Environmental Kuznets Curve

The Environmental Kuznets Curve (EKC) hypothesis modeling the relationship between economic growth and environment and suggesting that there is an inverted U-shape relationship between economic growth and environmental degradation is the model that has the largest area of application in the literature. The Environmental Kuznets Curve hypothesis also represents sustainable development very accurately. According to the EKC hypothesis, environmental degradation initially increases together with the increasing income up to a threshold value and once this threshold is passed, higher income value increases the environmental quality. The relationship reveals an inverted U-shape as shown in Fig. 1. The EKC hypothesis indicates a long-term relationship between environment and economic growth. As agricultural activities, extraction and usage of resources accelerate during the take-off stage of economic development, the resource consumption rate starts to exceed the resource renewal rate and therefore toxicity and amount of waste increase dramatically. On the other hand, when

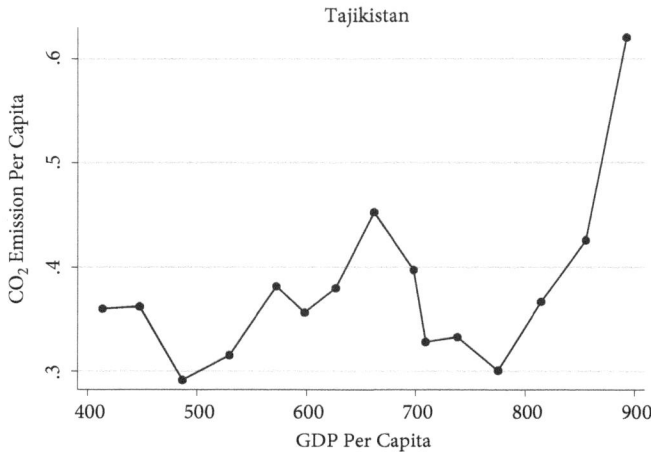

Fig. 2: Environmental Kuznets Curve (Low-Income Economies). Source: Created by Authors

we move on to higher development levels, structural changes in the knowledge-intensive industry and service sectors, increased environmental awareness, enforcement of environmental regulations, more environmentally friendly technologies and higher environmental expenditures first balance and then reduce environmental degradation (Dinda, 2004: 434).

Financial development is ignored in most studies analyzing the relationship between economic growth and environment. However, the role of financial markets in economies should not be underestimated. There are important reasons why the impact of financial development on environmental degradation must be investigated. Firstly, financial development can cause faster economic growth and therefore impact the environmental quality by increasing foreign direct investments and R&D investments. On the other hand, financial development provides, especially, developing countries with opportunities to use new technologies. Thus, it can enable such countries to make clean and environmentally friendly production. It is important to examine the effects of economic growth, energy consumption, commercial liberalization, urbanization and foreign direct investment as well as the impact of financial development on environmental degradation in Eurasian countries, which have undergone significant developments both economically and financially in recent years.

Fig. 1 shows a typical EKC curve. Fig. 2, 3 and 4 show the relationship between CO_2 emission and income per capita in Eurasian countries included in this study

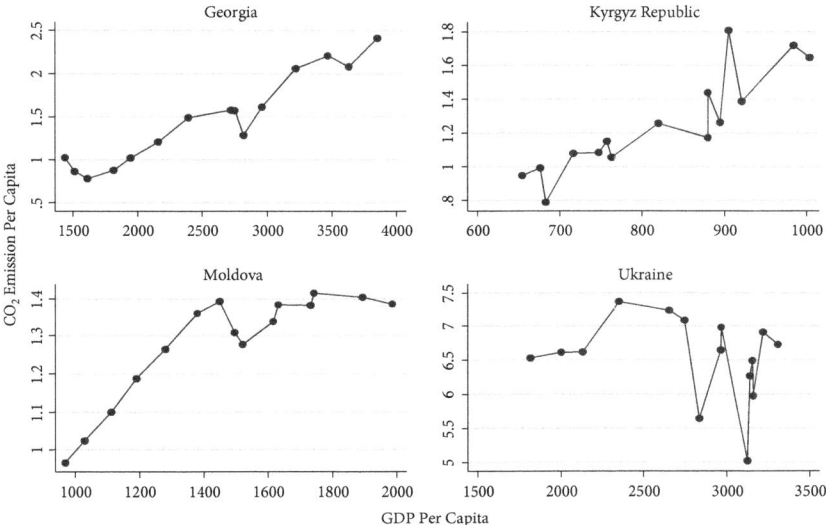

Fig. 3: Environmental Kuznets Curve (Lower-Middle-Income Economies). Source: Created by Authors

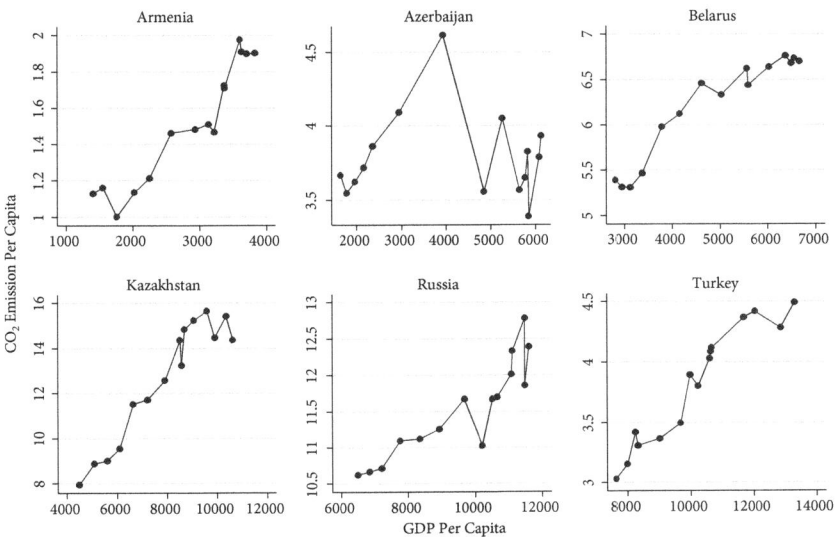

Fig. 4: Environmental Kuznets Curve (Upper-Middle-Income Economies). Source: Created by Authors

based on the country classification made by the World Bank according to income level. Graphics showing CO_2 emissions and per capita income of countries reveal that countries are a part of the inverted U-shape relationship in parallel with the economic development stage. Although the CO_2 emission decreased after 2008 in Tajikistan which is the only country in the low-income group, it started to increase again after 2010. Accordingly, CO_2 emission per capita increases as income increases. Hence, Tajikistan is in the take-off stage of the economy according to the EKC hypothesis. Fig. 3 shows countries in the lower-middle income group. We can say that the graphics of Georgia and Kyrgyzstan have an increasing trend. Similarly, Moldova has an increasing trend despite the fall in the last year. Although Ukraine started to decrease after 2003, it has a fluctuating relationship in recent years. As a result, we can say that the economy is in the developing stage in countries included in the lower-middle income group.

We can say that the countries in the upper-middle-income group (Fig. 4) have an increasing trend in general. Although countries other than Azerbaijan seem to have reached a turning point regarding the relationship between CO_2 emission and GDP per capita, further data are needed to confirm this assertion. Azerbaijan, on the other hand, follows a wavy pattern as we seen in Fig. 4. The possible cause for the decrease in the CO_2 emission may be the use of more efficient technologies. In general, it is hard to say that the EKC hypothesis regarding the Eurasian countries has reached a turning point. These countries generally exhibit an EKC pattern of an economy that is in the take-off stage. In addition to per capita income, factors such as the industrial structure peculiar to each economy, energy composition, technological level, commerce etc. cause different figures to emerge. This study analyzes the impacts of these factors.

This study aims to analyze the relationship among CO_2 emission, economic growth, energy consumption, financial development, trade liberalization and urbanization within the scope of the EKC hypothesis. Although there are many studies on the relationship between economic growth and environment, the number of studies on the role of financial development is limited. Particularly, we have not encountered any studies testing the relationship between such variables by taking cross-sectional dependence and coefficient homogeneity into account for Eurasian countries.

2 Literature Review

The relationship between economic growth and environment has been analyzed in many studies and it has become the center of attention for researches in the last twenty years. This relationship has been analyzed within the scope of the

EKC hypothesis in general. This hypothesis suggests an inverted U-shape relationship between economic growth and environmental quality. In other words, the damage to the environment increases until a threshold value is reached and then it starts to decrease. This curve was initially used to show the relationship between income and inequality as mentioned by Kuznets (1955). Later on, the relationship between long-term economic growth and the environment has been analyzed with reference to this model. Grossman and Krueger (1991) proposed this assumption for the first time. Following this study, Shafik and Bandyopadhyay (1992), Panayotou (1993), Cropper and Griffiths (1994), Selden and Song (1994), Dinda et al. (2000) and Stern and Common (2001) analyzed the environmental Kuznets curve using data related to environmental pollution and economic growth. Richmond and Kaufman (2006), Ang (2007), Soytaş et al. (2007), Zhang and Cheng (2009), Halıcıoğlu (2009), Apergis and Payne (2010), Öztürk and Acaravcı (2010), Acaravcı and Öztürk (2010), Pao and Tsai (2011), Öztürk and Uddin (2012), Saboori et al. (2012), Cho et al. (2014) and Bölük and Mert (2015) tested the relationship among environmental pollution, economic growth and energy within the scope of the EKC hypothesis.

Recent studies include variables such as trade liberalization, urbanization, population, financial development and globalization etc. in addition to economic growth and energy consumption variables within the framework of the EKC hypothesis in order to analyze their impacts on environmental pollution. Furthermore, some studies use data such as deforestation and ecological footprint instead of CO_2 emission as the pollution data. For instance, Ahmet et al. (2015) tested the validity of the EKC hypothesis for Pakistan by examining the deforestation, economic growth, energy consumption, trade openness and population data related to the period between 1980 and 2013 with ARDL Bound Test and found out that the hypothesis was valid. Al-Mulali et al. (2015) tested the validity of EKC between 1980 and 2008 using ecological footprint, growth, energy consumption, urbanization and trade openness data of 93 countries. The test results confirmed the validity of the hypothesis in upper-middle and high-income countries.

Farhani et al. (2013) analyzed the relationship among CO_2, economic growth, energy consumption, trade liberalization and urbanization within the framework of the EKC hypothesis for MENA countries for the period 1980–2009. The analysis results confirm the validity of the EKC hypothesis in this country group. Kasman and Duman (2015) used CO_2, economic growth, energy consumption, trade liberalization and urbanization data to test the EKC hypothesis for EU counties between 1992 and 2010 and they detected an inverted U-shape relationship. Li et al. (2016) analyzed the relationship among CO_2, economic growth, energy consumption, trade liberalization and urbanization for 28 Chinese states between 1996 and 2012 using the Generalized Method of Moments (GMM) and Autoregressive Distributed Lag (ARDL) methods. The analysis results confirm the validity of the EKC hypothesis.

The number of studies analyzing the impact of financial development on environmental degradation has increased in recent years. Tamazian et al. (2009) analyzed the impact of financial development on environmental degradation for the USA, Japan and BRIC countries between 1992 and 2004 within the framework of the EKC hypothesis and found out that financial development was a means of decreasing environmental degradation. Jalil and Feridun (2011) tested the impacts of economic growth, energy consumption and financial development for China between 1956 and 2003 using the ARDL method. The analysis results confirm the EKC hypothesis and show that financial development reduces financial degradation. Shahbaz et al. (2013) analyzed the relationship among environmental pollution, economic growth, coal consumption, trade liberalization and financial development for South Africa between 1965 and 2008 using the ARDL method. The results of the analysis confirm the validity of the EKC hypothesis and show that financial development reduces financial pollution. Javid and Sharif (2016) analyzed the relationship among CO_2 emission, per capita income, square of per capita income, trade liberalization and financial development for Pakistan between 1972 and 2013. According to the results, the EKC hypothesis is valid but financial development increases environmental pollution.

3 Model, Dataset and Methodology

Environmental degradation on a global scale has caused public concerns about environmental issues to increase and resulted in efforts aimed at investigating and understanding the causes of environmental degradation. Environmental impact of economic growth is one of the subjects that have been discussed most by economists in recent years. Especially, the connection between economic growth and the environment caused many debates in the 1990s and the studies in the literature on the relationship between pollution and income growth increased significantly (Dinda, 2004: 432). The common point of these studies is that economic quality is degraded in the early stages of economic growth while it starts to heal in the stages after economic development is achieved. In other words, environmental pressures increase faster than income in early stages while environmental pressures decrease compared to GDP during the development stage and at the high-income level. This systematic relationship between

Tab. 1: Variables and the Source of Dataset. Source: Created by Authors

Variable	Indicator	Unit of Measurement	Source
Carbon Dioxide Emission	CO_2	Per Capita Metric Ton	WDI
Energy Consumption	E	Per Capita Oil Equivalent (in kg)	WDI
GDP Per Capita	GDP	Constant 2010 US Dollar	WDI
Financial Development	FD	% of GDP	WDI
Index of Openness	TR	% of GDP	WDI
Urbanization	URB	% of Total Population	WDI

Note: FD, TR and URB indicate the share of sector-specific domestic credits in the GDP, the share of export and import in the GDP, and the share of urban population in the total population respectively

income change and environmental quality is called the Environmental Kuznets Curve. This inverted U-shape relationship gets its name from Kuznet's (1955) work suggesting a similar relationship between income inequality and economic development. The first empirical EKC studies were conducted by Grossman and Krueger (1995), Shafik and Bandyopadhyay (1992), Panayotou (1993) and Selden Song (1994). Many other studies have been published in the literature following these studies. In line with the studies analyzing the impact of economic growth on environmental degradation, this mode has been formulated as follows:

$$CO_{2it} = \beta_0 + \beta_1 E_{it} + \beta_2 GDP_{it} + \beta_3 GDP_{it}^2 + \beta_4 FD_{it} + \beta_5 TR_{it} + \beta_6 URB_{it} + \varepsilon_{it} \quad (1)$$

In the equation (1); CO_2, E, GDP, GDP^2, FD, TR and URB represent carbon dioxide emission, energy consumption, GDP per capita, square of GDP per capita, financial development, openness index and urbanization respectively. Tab. 1 provides brief information about the variables used in the study as well as the dataset sources. All variables have been converted into logarithmic forms.

The data in the study is used on an annual basis and covers the period between 2000 and 2014. The data was obtained from the World Development Indicators database of the World Bank. Descriptive statistics and relationship coefficients of the dataset are shown in Tab. 2 and Tab. 3. There is a high positive and statistically significant relationship among CO_2 emission per capita, GDP per capita and energy consumption. There is a low but negative and statistically significant relationship between the openness index and CO_2 emission. The study includes 11 Eurasian countries whose data could be accessed. The countries included in the study are Azerbaijan, Kazakhstan, Kyrgyzstan, Turkey, Armenia, Belarus, Georgia, Moldova, Russia, Tajikistan and Ukraine.

Tab. 2: Descriptive Statistics. Source: Created by Authors

Variable	(1) Mean	(2) SD	(3) Min	(4) Max
CO_2	4.577	4.138	0.292	15.647
E	1843.003	1411.097	279.197	5167.012
GDP	4343.213	3485.8	413.586	13312.02
FD	26.133	18.029	0.0008	90.573
TR	92.004	31.185	42.0	199.675
URB	55.831	14.916	26.4	76.277

Tab. 3: Pairwise Correlations. Source: Created by Authors

	CO_2	E	GDP	FD	TR	URB
CO_2	1.00					
E	0.969*	1.00				
GDP	0.716*	0.686*	1.00			
FD	0.391*	0.411*	0.420*	1.00		
TR	-0.260*	-0.218*	-0.591*	-0.157*	1.00	
URB	0.545*	0.640*	0.644*	0.366*	-0.349*	1.00

* indicates 5 % significance level.

Due to the increasing economic and financial integration and high level of globalization in recent years, using panel data methods that ignore cross-sectional dependence can yield unreliable results. Therefore, this study firstly investigates the presence of cross-sectional dependence between countries using the LM test of Breusch and Pagan (1980), CDLM and CE tests of Pesaran (2004) and LMadj test of Pesaran et al. (2008) Moreover, the delta tests developed by Pesaran and Yamagata (2008) were used to test the homogeneity of slope coefficients. The delta tests developed by Pesaran and Yamagata (2008) test the slope homogeneity of slope coefficients in linear panel data analyses. The delta tests were based on changes in the distribution statistics of Swamy. Two different test statistics are calculated to test the homogeneity of series:

$$\tilde{\Delta} = \sqrt{N}\left(\frac{N^{-1}\tilde{S}-k}{\sqrt{2k}}\right) \quad (2)$$

$$\tilde{\Delta}_{adj} = \sqrt{N}\left(\frac{N^{-1}\tilde{S}-E(\tilde{z}_{it})}{\sqrt{var(\tilde{z}_{it})}}\right) \quad (3)$$

N, S and k indicate the cross section number, Swamy test statistics and explanatory variable number respectively.

The Panel AMG (Augmented Mean Group) test developed by Eberhardt and Bond (2009) was used to estimate long-term co-integration coefficients. The Panel AMG test takes the cross-sectional dependence and coefficient heterogeneity between countries into account. Another advantage of using this methodology is that it enables monitoring parameters of non-stationary variables. For this reason, no preliminary test procedure (unit root or co-integration) is necessary to use this approach. In the first step of the test procedure, the main panel model (equation 4) is estimated with the first difference form and T-1 period dummy as follows:

$$\Delta CO_{2it} = \alpha_1 \Delta E_{it} + \alpha_2 \Delta GDP_{it} + \alpha_3 \Delta GDP_{it}^2 + \alpha_4 \Delta FD_{it} \\ + \alpha_5 \Delta TR_{it} + \alpha_6 \Delta URB_{it} + \sum_{t=2}^{T} p_t (\Delta D_t) + \mu_{it} \quad (4)$$

Here ΔD_t and P_t represent the first difference of T-1 dummy and parameter of the period dummies respectively. In the second step, the estimated P_t parameters are converted into φ_t showing the collective dynamic process as follows:

$$\Delta CO_{2it} = \alpha_1 \Delta E_{it} + \alpha_2 \Delta GDP_{it} + \alpha_3 \Delta GDP_{it}^2 + \alpha_4 \Delta FD_{it} \\ + \alpha_5 \Delta TR_{it} + \alpha_6 \Delta URB_{it} + d_i \varphi_t + \mu_{it} \quad (5)$$

$$\Delta CO_{2it} - \varphi_t = \alpha_1 \Delta E_{it} + \alpha_2 \Delta GDP_{it} + \alpha_3 \Delta GDP_{it}^2 + \alpha_4 \Delta FD_{it} \\ + \alpha_5 \Delta TR_{it} + \alpha_6 \Delta URB_{it} + \mu_{it} \quad (6)$$

First, the regression model specific to the group is adapted with φ_t and then the average values of the group-specific model parameters are calculated. For instance, the energy consumption parameter (α_1) is calculated as follows: $\alpha_{1,AMG} = 1/N \sum_{i=1}^{N} \alpha_{1,i}$.

4 Empirical Results

In this study, the relationship between environmental pollution, economic growth, energy consumption, commercial liberalization, urbanization and financial development was analyzed for Eurasian countries. For the first stage, the cross-sectional dependence and coefficient homogeneity were analyzed and the results are given in Tab. 4. According to the analysis results, the null hypothesis suggesting that there is no dependence between countries has been rejected in all tests. This means that a shock in one country within the group can affect the

Tab. 4: Cross Sectional Dependence and Coefficient Homogeneity Test Results. Source: Created by Authors

	Test Statistics	Probability Value
Cross-Sectional Dependence		
LM	415.653***	0.000
CD_{LM}	34.387***	0.000
CD	19.974***	0.000
LM_{adj}	24.133***	0.000
Homogeneity		
$\check{\Delta}$	25.159***	0.000
$\check{\Delta}_{adj}$	32.785***	0.000

*** indicates 1 % significance

other countries. Furthermore, results of the homogeneity test indicate country-specific heterogeneity between countries.

In the second stage of the analysis, because of the cross-sectional dependence and country-specific heterogeneity among the countries, the panel AMG coefficient estimator is used. Tab. 5 shows the panel and cross-section results of Panel AMG coefficient estimator. The panel results revealed that financial development in Eurasian countries reduced environmental pollution. Specifically, a 1 % increase in financial development reduces CO_2 emission by 0.32 %. According to the panel AMG results, the CKE hypothesis is valid in Eurasian countries. In addition, a 1 % increase in energy consumption increases CO_2 emission by 0.97 %. The impacts of trade liberalization and urbanization on environmental pollution were found out to be statistically insignificant. Country-specific results show that financial development coefficients are negative and statistically significant only for Azerbaijan and Moldova. When we examine the impacts of energy consumption on carbon dioxide in a country-specific manner, we observe that they are positive and statistically significant for all countries except Moldova. This indicates intense consumption of fossil-based energy in Eurasian countries. The results confirm the presence of the EKC hypothesis in Azerbaijan, Belarus, Georgia and Moldova. An inverted U-shape relationship is detected between environmental pollution and national income in Russia. According to trade liberalization results, it reduces environmental degradation only in Azerbaijan. On the other hand, it is determined that trade liberalization increased the CO_2 emission in Armenia and Moldova. The urbanization coefficient is significant and negative only in Russia.

Tab. 5: Panel AMG Results. Source: Created by Authors

	E	GDP	GDP^2	FD	TR	URB
Azerbaijan	1.084*** [0.287]	9.171** [4.423]	-0.557** [0.275]	-0.174** [0.084]	-0.397* [0.216]	2.712 [3.188]
Kazakhstan	0.600* [0.339]	8.232 [9.994]	-0.474 [0.566]	0.032 [0.101]	0.121 [0.338]	-10.424 [9.930]
Kyrgyzstan	1.239*** [0.180]	7.043 [20.663]	-0.532 [1.539]	0.099 [0.067]	0.006 [0.148]	1.587 [14.120]
Turkey	0.979*** [0.159]	1.931 [3.503]	-0.099 [0.189]	0.038 [0.053]	-0.033 [0.087]	-0.930 [0.978]
Armenia	1.287*** [0.236]	-4.584 [2.896]	0.298 [0.191]	-0.041 [0.041]	0.251* [0.150]	0.050 [4.912]
Belarus	0.133 [0.191]	4.406* [2.468]	0.242* [0.145]	0.002 [0.002]	-0.000 [0.066]	-0.405 [1.306]
Georgia	1.440** [0.698]	26.051** [11.641]	-1.662** [0.770]	-0.015 [0.213]	-0.504 [0.609]	19.787 [22.325]
Moldova	-0.042 [0.215]	19.319*** [3.994]	-1.276*** [0.268]	-0.270*** [0.102]	0.246** [0.121]	-6.782 [5.859]
Russia	0.974*** [0.130]	-3.017* [1.466]	0.175** [0.080]	-0.062 [0.042]	-0.021 [0.057]	-7.208** [2.906]
Tajikistan	1.687*** [0.329]	27.217 [50.967]	-2.102 [4.028]	0.046 [0.059]	-0.047 [0.226]	69.948 [101.968]
Ukraine	1.384*** [0.213]	0.998 [5.285]	-0.072 [0.337]	-0.008 [0.041]	-0.191 [0.147]	2.074 [2.364]
Panel	0.979*** [0.163]	8.797*** [3.298]	-0.595*** [0.232]	-0.320*** [0.032]	-0.051 [0.071]	6.400 [6.779]

***, ** , and * indicate 1 %, 5 % and 10 % significance respectively. The values in brackets are standard errors

Conclusion

This study aims to analyze the relationship among CO_2 emission, economic growth, energy consumption, financial development, trade liberalization and urbanization for eleven Eurasian countries (Azerbaijan, Kazakhstan, Kyrgyzstan, Turkey, Armenia, Belarus, Georgia, Moldova, Russia, Tajikistan and Ukraine) for the period from 2000–2014 within the framework of the EKC hypothesis. Therefore, the EKC hypothesis and Panel AMG coefficient estimator are used for the analysis. Firstly, cross-sectional dependence and coefficient homogeneity tests are used in order to use the panel AMG test. The results show country-specific dependence and country-specific heterogeneity.

According to the Panel AMG results, financial development reduces CO_2 emission in Eurasian countries. The Panel results indicate the validity of the EKC hypothesis. Furthermore, it is found out that energy consumption increased environmental pollution while trade liberalization and urbanization coefficients were statistically insignificant. Based on the results of this study, it is important to benefit from policies that encourage renewable energy and energy efficiency and assist in the implementation of green energy in Eurasian countries. Moreover, the increase in energy efficiency and the role of renewable energy in total energy consumption can help decrease the dependence of this region on imported fossil fuels and increase energy security. The analysis results show that the financial development coefficient is negative, financial development does not happen at the cost of environmental pollution in Eurasian countries and it can help to procure investment capital to build new environmental facilities.

References

Acaravcı, A. and Öztürk, İ. (2010) "On the Relationship between Energy Consumption, CO_2 Emissions and Economic Growth in Europe", Energy, 35(12): 5412–5420.

Ahmed, K., Shahbaz, M., Qasim, A. and Long, W. (2015) "The Linkages between Deforestation, Energy and Growth for Environmental Degradation in Pakistan", Ecological Indicators, 49: 95–103.

Al-Mulali, U., Weng-Wai, C., Sheau-Ting, L. and Mohammed, A. H. (2015) "Investigating the Environmental Kuznets Curve (EKC) Hypothesis by Utilizing the Ecological Footprint as an Indicator of Environmental Degradation", Ecological Indicators, 48: 315–323.

Ang, J. B. (2007) "CO_2 Emissions, Energy Consumption, and Output in France", Energy Policy, 35(10): 4772–4778.

Apergis, N. and Payne, J. E. (2009) "Energy Consumption and Economic Growth in Central America: Evidence from a Panel Cointegration and Error Correction Model", Energy Econ., 31: 211–216.

Bölük, G. and Mert, M. (2015) "The Renewable Energy, Growth and Environmental Kuznets Curve in Turkey: An ARDL Approach", Renewable and Sustainable Energy Reviews, 52: 587–595.

Breusch, T. S. and Pagan, A. R. (1980) "The Lagrange Multiplier Test and its Applications to Model Specification in Econometrics", The Review of Economic Studies, 47(1): 239–253.

Cho, C. H., Chu, Y. P. ve Yang, H. Y. (2014) "An Environment Kuznets Curve for GHG Emissions: A Panel Cointegration Analysis", Energy Sources, Part B: Economics, Planning, and Policy, 9(2): 120–129.

Cropper, M. ve Griffiths, C. (1994) "The Interaction of Population Growth and Environmental Quality", The American Economic Review, 84(2): 250–254.

Dinda, S. (2004) "Environmental Kuznets Curve Hypothesis: A Survey", Ecological Economics, 49(4): 431–455.

Dinda, S., Coondoo, D. ve Pal, M. (2000) "Air Quality and Economic Growth: An Empirical Study", Ecological Economics, 34(3): 409–423.

Eberhardt, M. and Bond, S. (2009) "Cross-Section Dependence in Nonstationary Panel Models: a Novel Estimator", MPRA Paper 17692, University Library of Munich, https://mpra.ub.uni-muenchen.de/17692/ Access Date: 10.01.2019

Farhani, S., Shahbaz, M. and Arouri, M. E. H. (2013) "Panel Analysis of CO_2 Emissions, GDP, Energy Consumption, Trade Openness and Urbanization for MENA Countries", Munich Personal RePEc Archive, 1-19. https://mpra.ub.uni-muenchen.de/49258/1/MPRA_paper_49258.pdf Access Date: 10.01.2019

Grossman, G. M. ve Krueger, A. B. (1991) "Environmental Impacts of a North American Free Trade Agreement", National Bureau of Economic Research, (No. w3914).

Halicioglu, F. (2009) "An Econometric Study of CO_2 Emissions, Energy Consumption, Income and Foreign Trade in Turkey", Energy Policy, 37(3): 1156–1164.

Jalil, A. and Feridun, M. (2011) "The Impact of Growth, Energy and Financial Development on the Environment in China: A Cointegration Analysis", Energy Economics, 33(2): 284–291.

Javid, M. and Sharif, F. (2016) "Environmental Kuznets Curve and Financial Development in Pakistan", Renewable and Sustainable Energy Reviews, 54: 406–414.

Kasman, A. and Duman, Y. S. (2015) "CO_2 Emissions, Economic Growth, Energy Consumption, Trade and Urbanization in New EU Member and Candidate Countries: A Panel Data Analysis", Economic Modelling, 44: 97–103.

Kuznets, S. (1955). "Economic Growth and Income Inequality", The American Economic Review, 45(1), 1–28.

Li, T., Wang, Y. and Zhao, D. (2016) "Environmental Kuznets Curve in China: New Evidence from Dynamic Panel Analysis", Energy Policy, 91: 138–147.

Öztürk, I. and Acaravcı, A. (2010) "CO_2 Emissions, Energy Consumption and Economic Growth in Turkey", Renewable and Sustainable Energy Reviews, 14(9): 3220–3225.

Öztürk, İ. and Salah Uddin, G. (2012) "Causality among Carbon Emissions, Energy Consumption and Growth in India", Ekonomska Istraživanja, 25(3): 752–775.

Panayotou, T. (1993) "Empirical Tests and Policy Analysis of Environmental Degradation at Different Stages of Economic Development", International Labour Organization, No. 992927783402676.

Pao, H. T., and Tsai, C. M. (2011). "Modeling and Forecasting the CO_2 Emissions, Energy Consumption, and Economic Growth in Brazil", Energy, 36(5): 2450–2458.

Pesaran, M. H. (2004) "General Diagnostic Tests for Cross Section Dependence in Panels" Cambridge Working Papers in Economics 0435, Faculty of Economics, University of Cambridge.

Pesaran, M. H., Ullah, A. and Yamagata, T. (2008) "A Bias-Adjusted LM Test of Error Cross-Section Independence", The Econometrics Journal, 11(1): 105–127.

Pesaran, M. H. and Yamagata, T. (2008) "Testing Slope Homogeneity in Large Panels", Journal of Econometrics, 142(1): 50–93.

Richmond, A. K. and Kaufmann, R. K. (2006) "Is there a Turning Point in the Relationship between Income and Energy Use and/or Carbon Emissions", Ecol. Econ., 56: 176–189.

Saboori, B., Sulaiman, J. and Mohd, S. (2012) "Economic Growth and CO_2 Emissions in Malaysia: A Cointegration Analysis of the Environmental Kuznets Curve", Energy Policy, 51: 184–191.

Selden, T. M. and Song, D. (1994) "Environmental Quality and Development: Is there a Kuznets Curve for Air Pollution Emissions?", Journal of Environmental Economics and Management, 27(2): 147–162.

Shafik, N. and Bandyopadhyay, S. (1992) "Economic Growth and Environmental Quality: Time Series and Cross-Country Evidence", Policy Research Working Paper Series 904, The World Bank.

Shahbaz, M., Tiwari, A. K. and Nasir, M. (2013) "The Effects of Financial Development, Economic Growth, Coal Consumption and Trade Openness on CO_2 Emissions in South Africa", Energy Policy, 61: 1452–1459.

Soytaş, U., Sarı, U. and Ewing, B.T. (2007) "Energy Consumption, Income and Carbon Emissions in the United States", Ecological Economics, 62: 482–489.

Stern, D. I. and Common, M. S. (2001) "Is there an Environmental Kuznets Curve for Sulfur?", Journal of Environmental Economics and Management, 41(2): 162–178.

Tamazian, A., Chousa, J. P. and Vadlamannati, K. C. (2009) "Does Higher Economic and Financial Development Lead to Environmental Degradation: Evidence from BRIC Countries", Energy Policy, 37(1): 246–253.

Yandle B., Bhattarai M., and Vijayaraghavan M. (2004) Environmental Kuznets curves: a review of findings, methods, and policy implications, PERC Research Study 02-1, Bozeman, MT: Property and Environmental Research Center.

Zhang, X.P. and Cheng, X.M. (2009) "Energy Consumption, Carbon Emissions and Economic Growth in China", Ecological Economics, 68: 2706–2712.

Celil Aydın, Burak Darıcı and Şeyma Şahin Kutlu

Economic Growth and Ecological Footprint: Reconsidering the Empirical Basis of Environmental Kuznets Curves

Introduction

The main goal of economic activities is to increase welfare by meeting the requests and needs of people. Therefore, increasing production and income is the ultimate goal of economies. Countries initially focus on increasing their revenues and ignore environmental problems such as environmental pollution, climate change and global warming etc. This increases the pressure on the environment and natural resources by polluting the air through the increased carbon density in the atmosphere, the nature through accumulation of non-recyclable waste and water through increased industrial and domestic waste. Sustainability of an economic-growth mentality that only focuses on the increase in production and does not take environmental conditions into account, has started to be questioned. At this point, policy makers and researchers have started to find ways to increase production without destructing the environment and survive without exceeding the renewal capacity of natural resources along with economic growth. In this context, a growth mentality in which market economy is not subject to any limitations has given way to a sustainable development model and new production models that do not damage the environment have been tried.

A growth mentality serving only the market economy dominated the industrialization policies until 1960. The industrialization policies at that time were based on the notion that the physical conditions in the world would be enough to enable all economies to be industrialized and increase the existing industrialization capacity. Another prevailing opinion was that the nature would not remain polluted and it would renew itself (Karabıçak and Armağan, 2004: 204). The rapid growth of the world population increased economic activities in the period following 1960. This upset the balance between resources and needs by causing natural resources, which were regarded as free goods, to be consumed excessively. Furthermore, the limited nature of environmental problems carried negative externalities beyond national borders. At this point, it was pointed out that people worsened the environmental conditions and people started to think that those problems were actually economic problems rather than being environmental problems. An extensive literature emerged in the field of economy about the environment and environmental problems which had been mentioned in various platforms before. Accordingly, the report titled Limits to Growth was published under the leadership of the Club of Rome. This report stated that environmental degradation increased in parallel with economic growth, transformed into an effective paradigm in time and drew attention to the industrialization-environment dilemma. Tangible suggestions were made for this dilemma afterwards. On the other hand, the findings of the studies conducted by Grossman and Krueger (1991, 1995), Shafik and Bandyopadhyay (1992), Panayotou (1993) and Selden and Song (1994) caused the Limits to Growth paradigm to be questioned. This new argument is called Environmental Kuznets Curve hypothesis as it is an extension of the study conducted by Kuznets. According to this hypothesis, environmental pollution increases in the early stages of economic growth, but it decreases in the following stages. This hypothesis suggests an inverted U correlation between economic growth and environmental pollution. In other words, this hypothesis is based on the opinion that individuals prioritize their own needs at low income levels whereas environmental awareness increases in parallel with the increase in the income level (Albayrak and Gökçe, 2015: 286). These findings of the EKC hypothesis are frequently discussed and questioned at the theoretical and empirical level. The first group of criticism is related to the shape of the EKC curve (Agras and Chapman, 1999; Harbaugh et al., 2002; Roca, 2003; Stern, 2004; Dinda, 2004; Richmond and Kaufmann, 2006; Fürstenberger and Wagner, 2007; Wagner, 2008; Vollebergh et al., 2009; Kearsley and Riddel, 2010; Chiu, 2012; Herranz and Lorente, 2016; Stern, 2017). This creates an expectation of a flatter EKC (Dasgupta et al., 2002: 152). The second group of criticism is related to the idea that the EKC hypothesis can be valid at a local and regional level but its validity at the global level could be irrelevant (Gill et al., 2018). Findings obtained from empirical studies vary depending on the selected environmental quality type, the analyzed country, time period, econometric method and other explanatory variables. Results of some studies support the EKC hypothesis (Ang, 2007; Jalil-Mahmud, 2009; Lean and Smyth, 2010; Güriş and Tuna, 2011; Nasir and Rehman, 2011; Shahbaz et al., 2012; Erataş and Uysal, 2014; Ergün and Polat, 2015; Kasman and Duman, 2015; Li et al., 2016; Aytun et al., 2017; Moutinho et al., 2017) while other studies (Richmond and Kaufmann, 2006; Karaca, 2012; Sarısoy and Yıldız, 2013; Herranz and Lorente, 2016) did not find any correlations. Many pollutants that emerge in parallel with the growth and development of economies cause significant environmental problems at the local and global level. The demand for natural resources has reached an unsustainable level. In order to draw attention to these problems, the ecological footprint concept used to measure the environmental sustainability was developed in the 1990s (Özsoy, 2015: 201). This concept is a significant indicator that represents reproduction of consumed resources with current technology and resource management and the size of land and water required to dispose of the waste resulting from such activities. This new concept indicates the magnitude and source of the pressure created on natural resources due to production and consumption activities so that the impacts of humans on the environment can be evaluated. This concept has provided a new point of view for the correlation between the nature and humans and the resulting calculations constitute one of the important steps of progress into the future (Altıparmak and Avcı, 2011: 38). Although the relevant literature is considerably limited, the number of studies has increased after the pioneering studies conducted by Rees and Wackernagel (1994). Some of the studies analyzing this subject in terms of ecological footprint (York et al., 2004; Bagliani et al., 2008; Caviglia-Harris et al., 2009; Mostafa, 2010; Wang et al., 2013) could not obtain results supporting the EKC hypothesis while other studies (Al-Mulali et al., 2015; Charfeddine and Mrabet, 2017) obtained results that supported the EKC hypothesis.

Most studies are based on parametric functional forms. As such studies can be insufficient for determining the correlation between the environment and income, researchers started to use non-linear models that have threshold effects but do not have any parametric correlations. In this sense, this study analyzes the validity of the EKC hypothesis for 80 countries in the period between 1975 and 2013 as well as the role of real income in the non-linear correlation between ecological footprint and real income with the help of smooth transition regression models.

1 Methodology

This study analyzes the non-linear correlation between ecological footprint and per capita income using the non-linear panel data analysis method. The first method analyzing the non-linear correlation between variables in the panel data analysis is the Panel Threshold Regression (PTR) method developed by Hansen (1999). In this method, the impact of the threshold variable on the dependent variable differ in regimes below and above the threshold. This requires the coefficient showing the impact of the threshold variable on the dependent variable to differ across regimes.

Parameters are assumed to change suddenly between regimes in the PTR approach and each regime is distinguished from the others according to the determined threshold values. However, the sudden parameter change among regimes is not always possible in terms of economics (Güloğlu and Nazlıoğlu, 2013: 11). In the correlation between ecological footprint and per capita income, this approach divides countries included in the panel into groups of per capita income and estimates different parameters for each group. As a result, it is assumed that there are clear differences between developed countries with high per capita income and developing countries with low per capita income. So it is accepted that a developing country suddenly becomes a developed country. But transition from a developing country into a developed country takes time. This means that the estimated parameters change slowly, not suddenly. Therefore, the Panel Smooth Transition Regression (PSTR) model developed by Gonzalez, Terasvirta and van Dijk (2005) that allows regression parameters to change gradually and slowly rather than changing suddenly while moving from one regime to another was used.

A fixed PSTR model with two regimes was created in order to analyze the non-linear correlation between ecological footprint and per capita income and this model is shown in equation (1):

$$LnEF_{i,t} = \mu_i + \beta_0 LnGDP_{i,t} + \beta_1 LnGDP_{i,t} * g(q_{i,t}; \gamma, \theta) + \varepsilon_{i,t} \quad (1)$$

Here, $LnEF$ represents log-transformed per capita ecological footprint; $LnGDP$ is log-transformed per capita real GDP; ε is the error term; $t = 1, 2..., T$ time periods and $i = 1, 2, 3..., N$ countries. Coefficients μ_i allow for the possibility of unit-specific fixed effects, and the variable q_i is a potential threshold variable. In equation (1), μ_i allow for the possibility of unit-specific fixed effects, and the variable q_i is a potential threshold variable. In equation (1), $g(q_{i,t}; \gamma, \theta)$ is used as a transition function and it is expressed as in equation (2) in the logistics function form:

$$g(q_{i,t}; \gamma, \theta) = \left[1 + \exp(-\gamma(q_{i,t} - \theta))\right]^{-1} \quad (2)$$

θ parameter in equation (2) is the threshold parameter between two regimes corresponding to $g(q_{i,t}; \gamma, \theta) = 0$ ve $g(q_{i,t}; \gamma, \theta) = 1$ while γ (smoothing parameter) represents the smoothness of the change in the value of the transition function, i.e. transition from one regime to another. As the smoothing parameter goes to infinity, change from 0 to 1 in the ($\gamma \to \infty$) transition function happens suddenly and sharply as in PTR model of transition from one regime to another at the point where the threshold variable is equal to θ. In this case, the model is estimated using the PTR approach. When the smoothing parameter approaches zero, $(\gamma \to 0)$ transition function equals a constant and the model is reduced to the linear form at the point where the smoothing parameter equals zero ($\gamma = 0$). In this case, the model is estimated using a cross-sectional estimator (Fouquau et al., 2008: 287–288).

The transition function is the continuous function of the transition variable and obtains a value between 0 and 1. When the transition function value is 0 $\left(g(q_{i,t};\gamma,\theta)=0\right)$ in equation (1), the value of the regression coefficient becomes β_0; when it is 1 $\left(g(q_{i,t};\gamma,\theta)=1\right)$, the value of the regression coefficient becomes $\beta_0 + \beta_1$. On the other hand, when the transition function assumes a value between 0 and 1 ($0 < g(q_{i,t};\gamma,\theta) < 1$), regression coefficient is calculated as the weighted average of β_0 and β_1. Therefore, it is preferred to interpret coefficient signs instead of interpreting coefficients directly in the PSTR model (Fouquau et al., 2008; 287–288). In other words, it is stated if the impact of the independent variable on the dependent variable is positive or negative and time-dependent flexibilities are interpreted (Güloğlu and Nazlıoğlu, 2013: 12).

The PSTR model can have two or more regimes. In this case, the PSTR model with more than two regimes related to the model in equation (1) is expressed as in equation (3):

$$LnCO_{2i,t} = \mu_i + \beta_0 LnGDP_{i,t} + \sum_{j=1}^{r} \beta_j LnGDP_{i,t} * g_j\left(q_{i,t}^{(j)};\gamma_j,\theta_j\right) + u_{i,t} \qquad (3)$$

The transition function used in the PSTR model with more than two regimes is expressed as in equation (4).

$$g(q_{i,t};\gamma,\theta) = \left[1 + \exp\left(-\gamma \prod_{j=1}^{m}(q_{i,t} - \theta_j)\right)\right]^{-1}, \quad \gamma > 0, c_1 \leq c_2 \leq \cdots \leq c_m \qquad (4)$$

In the event that the transition (threshold) variable is different from (q) explanatory variable $\left(q \neq LnGDP_{i,t}\right)$, the flexibility value is calculated as in equation (5):

$$e_{i,t} = \frac{\partial LnCO_{2i,t}}{\partial LnGDP_{i,t}} = \beta_0 + \sum_{j=1}^{r} \beta_j * g_j\left(q_{i,t}^{(j)};\gamma_j,\theta_j\right) \qquad (5)$$

In the event that the transition (threshold) variable is one of the (q) explanatory variables $\left(q = LnGDP_{i,t}\right)$, the flexibility value is calculated as in equation (6):

$$e_{i,t} = \frac{\partial LnCO_{2i,t}}{\partial LnGDP_{i,t}} = \beta_0 + \sum_{j=1}^{r} \beta_j * g_j\left(q_{i,t}^{(j)}; \gamma_j, \theta_j\right)$$
$$+ \sum_{j=1}^{r} \beta_j \frac{\partial g_j\left(q_{i,t}^{(j)}; \gamma_j, \theta_j\right)}{\partial LnGDP_{i,t}} LnGDP_{i,t} \quad (6)$$

PSTR analysis is conducted in three steps: testing linearity, determining the number of regimes (r) and estimation (Fouquau et al., 2008; 287–288). Linearity is determined by testing $\gamma = 0$ or $\beta_0 = \beta_1$ null hypotheses. But in both cases, the test statistics cannot be standard as the model has parameters that are not defined in the null hypothesis. For this reason, first degree Taylor expansion is performed for $\gamma = 0$ instead of a transition function. The null hypothesis in the linearity test is the linear model while the alternative hypothesis is the PSTR model. These hypotheses are tested with the standard F-statistic. Rejecting the null hypothesis according to the standard F-statistic makes it necessary to use the PSTR model. The number of regimes is determined after the linear model hypothesis is rejected. In this stage, firstly the r = r* = 1 null hypothesis is tested against the r = r* + 1 alternative hypothesis. The process ends if the null hypothesis is accepted. In the event that the null hypothesis is rejected, the r = r* + 1 null hypothesis is tested against the r = r* + 2 alternative hypothesis. The stage in which the number of regimes is determined continues until the null hypothesis is accepted for the first time (Fouquau et al., 2008; 287–288). The final stage of PSTR analysis is the estimation stage. In this stage, first the fixed impacts of cross sections constituting the panel are extracted from the time averages of variables and then the converted model is estimated with non-linear Ordinary Least Squares (Gonzalez et al., 2005).

2 Data, Empirical Results and Implications

2.1 Data Specifications

In this study, the relationship between ecological footprint and real income was investigated using panel smooth transition regression analysis which takes into account the per capita real GDP threshold level. This study uses annual and balanced panel data covering 80 countries for the period between 1975 and 2013 and further divides the dataset into advanced and non-advanced countries.[1]

---

[1] According to IMF country classification, we split our sample into two groups as advanced and non-advanced countries. Among these 80 countries, 24 countries were advanced countries, while 56 countries were non-advanced countries.

Tab. 1: Descriptive Statistics of Variables in Levels over the Period 1975–2013. Source: Created by Authors

Full Sample	Per Capita EF	Per Capita Real GDP
Mean	3.41	13932.91
Std. Dev.	2.61	18295.97
Max.	17.19	111069.20
Min.	0.43	169.33
Obs.	3120	3120
Advanced Countries		
Mean	6.51	34881.37
Std. Dev.	2.34	16514.16
Max.	17.19	111069.20
Min.	1.66	2840.01
Obs.	936	936
Non-advanced Countries		
Mean	2.09	4954.99
Std. Dev.	1.23	9626.13
Max.	9.85	94903.19
Min.	0.43	169.33
Obs.	2184	2184

Note: Std. Dev. is the abbreviation of standard deviation. Max. is the maximum value. Min. is the minimum value. Obs. means the number of observation.

In this study the variables include per capita ecological footprint ($LnEF$) and per capita real GDP ($LnGDP$), which are respectively measured as global hectares (gha) per person and constant 2010 US dollars. Per capita ecological footprint and real GDP data comes from the Global Footprint Network (GFN) and World Development Indicators (WDI) database respectively. All variables are expressed in natural logarithm. Tab. 1 reports the descriptive statistics of all variables.

As shown in Tab. 1, on average, per capita ecological footprint and per capita real GDP for 80 countries are approximately 3.41 gha and US$13,932.91; for advanced countries 6.51 gha and US$34,881.37; for non-advanced countries 2.09 gha and US$4,954.99, respectively. Tab. 1 also shows that, on average, advanced countries have greater values for all variables than non-advanced countries. In addition, per capita real GDP is significantly and highly positively correlated to per capita ecological footprint for full sample (0.85). This result means that countries with higher income levels emit higher ecological footprint.

Tab. 2: Cross Section Dependence. Source: Created by Authors

Full Sample	LnEF	LnGDP	Model
CD_{BP}	4642.866***	5631.447***	11857.508***
CD_{LM}	18.653***	31.088***	109.405***
CD	-1.544*	3.306***	18.013***
LM_{adj}	3.582***	6.665***	67.582***
Advanced Countries			
CD_{BP}	416.465***	469.113***	1825.931***
CD_{LM}	5.979***	8.219***	65.969***
CD	-2.095***	-2.751***	30.635***
LM_{adj}	-0.798*	3.154***	17.512***
Non-advanced Countries			
CD_{BP}	2370.581***	2779.798***	2480.167***
CD_{LM}	20.284***	28.230***	22.412***
CD	-3.875***	-3.774***	-3.509***
LM_{adj}	57.075***	40.296***	6.962***

*, **, *** indicate significance at 10 %, 5 % and 1 % levels, respectively. CD_{BP}: Breusch and Pagan 1980 test, CD_{LM}: Pesaran 2004 CDlm test, CD: Pesaran 2004 CD test and LM_{adj}: Bias-adjusted CD test.

2.2 Empirical Results

Firstly, the dependence between cross sections (countries) was analyzed in this study examining the non-linear correlation between per capital ecological footprint and per capita income. Whether the dependence between the cross sections constituting the panel is taken into account affects the estimation results significantly (Breusch and Pagan, 1980). Therefore, firstly the presence of cross section dependence in the series and model must be tested. Also this must be taken into account while selecting the unit root tests. Otherwise, analyses could produce faulty results. For this reason, first the presence of cross section dependence was analyzed using the Adjusted Lagrange Multiplier (LM_{adj}) that was developed by Breusch-Pagan (1980) and whose deviation was adjusted by Pesaran et al. (2008); the results are given in Tab. 2.

According to the test statistics related to per capita GDP and per capita ecological footprint series in Tab. 2, the hypothesis suggesting that there is no cross section dependence has been strongly rejected. It was decided that cross section dependence existed in the series and in the model. This result indicates that a shock in one country affects others as well. Furthermore, test methods taking the cross section dependence into account must be used while selecting

Tab. 3: Results of Moon and Perron's (2004) Panel Unit Root Tests. Source: Created by Authors

Full Sample	LnEF	LnGDP
$\breve{r}$	4	4
t_a^*	-25.025 (0.000)	-21.739 (0.000)
t_b^*	-11.756 (0.033)	-9.606 (0.038)
$\breve{\rho}_{pool}^*$	0.854	0.865
Advanced Countries		
$\breve{r}$	3	3
t_a^*	-10.209 (0.000)	-14.394 (0.000)
t_b^*	-5.661 (0.075)	-6.432 (0.063)
$\breve{\rho}_{pool}^*$	0.897	0.868
Non-advanced Countries		
$\breve{r}$	3	5
t_a^*	-15.718 (0.000)	-19.589 (0.000)
t_b^*	-8.534 (0.071)	-9.035 (0.008)
$\breve{\rho}_{pool}^*$	0.895	0.872

Notes: $\breve{r}$ is the estimated number of common factors. $\breve{r}$ is the estimated number of common factors. t_a^* and $\breve{r}$ is the estimated number of common factors. t_a^* and t_b^* are the unit root test statistics based on de-factored panel data. Corresponding *p-values* are in parentheses. $\breve{r}$ is the estimated number of common factors. t_a^* and t_b^* are the unit root test statistics based on de-factored panel data. Corresponding *p-values* are in parentheses. $\breve{\rho}_{pool}^*$ is the corrected pooled estimates of the autoregressive parameter.

the methods to be used in the following stages of the analysis. Therefore, the stationarity of the series is analyzed in the following sections using the second generation panel unit root test developed by Moon and Perron (2004) that takes cross section dependence into account. The results are given in Tab. 3. According to the results in Tab. 3, the null hypothesis created with the full sample stating that the series contain unit roots in non-advanced and advanced countries is rejected for per capita GDP and per capita ecological footprint. This indicates that the series are stationary at the level values (*I(0)*).

Tab. 4: Tests for the Linearity. Source: Created by Authors

Threshold variables (*LnGDP*)	Full Sample	Advanced Countries	Non-advanced Countries
H_0 : *Linear Model vs* H_1 : *PSTR Model at least one Threshold Variable*			
LM	8.784***	261.443***	37.471***
	(0.003)	(0.000)	(0.000)
LM_F	8.580***	176.348	37.130***
	(0.003)	(0.000)	(0.000)
LRT	8.796***	306.596	37.797***
	(0.003)	(0.000)	(0.000)

Notes: Under H_0, the *LM* and *LR* statistics have an asymptotic $\chi^2(mK)$ distribution, whereas LM_F has an asymptotic $F(mK, TN - N - m(K+1))$ distribution. Moreover, *r* is the number of transition functions. *P-values* are in parentheses. ** and *** indicates the 5 % and 1 % significance level respectively.

The stage after determining that the variables used in the model are stationary at the level values is the first stage of PSTR analysis in which the linear model is tested against the non-linear model. Tab. 4 shows the results of the Wald Tests (*LM*), Fisher Tests (LM_F) and LRT Tests (*LRT*) calculated with the full sample in order to test linearity in advanced and non-advanced countries and determine the number of transition functions.

As you can see in Tab. 4, the null hypothesis is rejected at 1 % significance level both for the full sample and for advanced and non-advanced countries according to the LM, LMF and LRT test results. This way, it was accepted as alternative hypothesis that the model included at least one non-linear threshold effect in all three groups and it was concluded that it would be suitable to use linear models to model the impact of per capita GDP on per capita ecological footprint. The next stage of the analysis after rejection of the linear model hypothesis for all three groups is determining the number of regimes. The *LM*, LM_F and *LRT* tests were repeated with the full sample for advanced and non-advanced countries in order to determine the suitable number of regimes and the results are given in Tab. 5.

As you can see in Tab. 5, the null hypothesis that the model had a threshold effect was rejected for the full sample but it was not rejected for advanced and non-advanced countries. This way, it was concluded that the model had a threshold effect in advanced and non-advanced countries and it would be estimated with two-regime PSTR model. On the other hand, rejecting the null hypothesis for

Tab. 5: Tests for the Remaining Non-Linearity of the PSTR Model. Source: Created by Authors

Threshold variables (LnGDP)	Full Sample	Advanced Countries	Non-advanced Countries
$H_0: r=1$ vs $H_1: r=2$			
LM	151.251***	0.859	0.510
	(0.000)	(0.651)	(0.475)
LM_F	154.729***	0.417	0.496
	(0.000)	(0.659)	(0.481)
LRT	155.041***	0.860	0.510
	(0.000)	(0.651)	(0.475)
$H_0: r=2$ vs $H_1: r=3$			
LM	1.315		
	(0.251)		
LM_F	1.280		
	(0.258)		
LRT	1.315		
	(0.251)		

Notes: Under H_0, the LM and LR statistics have an asymptotic $\chi^2(mK)$ distribution, whereas LM_F has an asymptotic $F(mK, TN - N - m(K+1))$ distribution. Moreover, r is the number of transition functions. *P-values* are in parentheses. ** and *** indicates the 5% and 1% significance level respectively.

the full sample and accepting the alternative hypothesis that the models have "at least two threshold effects" require the suitable regime number to be determined accurately. Therefore, the model was retested with the alternative hypothesis that it had "at least two threshold effects" against the null hypothesis that it had "two threshold effects" for the full sample. According to the results, the null hypothesis that the model had "two threshold effects" could not be rejected for the full sample. As a result, it was concluded that the model had two threshold effects and it would be estimated with three-regime PSTR models for the full sample. In the next stage, the non-linear correlation between per capita ecological footprint and per capita income was estimated with a two-regime PSTR model for advanced and non-advanced countries and with a three-regime PSTR model for the full sample, and the estimation results are given in Tab. 6.

The results in Tab. 6 indicate that the estimated slope parameters are relatively small in all models, ranging from 2.315 to 15.619 implying that a continuum of conditions among regimes occurs – that is, the relationship between ecological

Tab. 6: Estimated Results of the PSTR Model. Source: Created by Authors

Threshold Variables ($LnGDP$)	Full Sample	Advanced Countries	Non-Advanced Countries
$LnGDP_1$	0.385***	0.456***	0.381***
	(0.012)	(0.018)	(0.012)
$LnGDP_2$	0.059***	-0.091***	0.124***
	(0.005)	(0.005)	(0.009)
$LnGDP_3$	-0.07***		
	(0.004)		
Location parameters, θ	9.596	10.137	9.700
	10.865		
Slope parameters, γ	2.770	2.315	15.619
	11.277		

Notes: Standard errors are corrected for heteroskedasticity in parentheses. All variables are expressed in natural logarithm. *** indicates the 1 % significance level.

footprint and real income smoothly switches from one regime to another regime (see Figs. 1–3 in the Appendix for full sample, advanced and non-advanced countries).

As you can see in Tab. 6, the estimated coefficient of per capita GDP in the first regime of the models related to advanced and non-advanced countries as well as the full sample (β_0) is statistically significant and positive (0.385, 0.456 and 0.381 respectively). In the second regime, the coefficient indicated as the sum of β_0 and β_1 is still positive (0.44, 0.365 and 0.505 respectively) and it obtained a value higher than the first regime in the full sample and non-advanced countries while the value was lower in the advanced countries. In the full sample with the third regime, the coefficient is expressed as the sum of β_0, β_1 and β_2 and its value is lower than the first and second regime but still positive (0.374).

The results indicate that an increase in real income increases the ecological footprint in the full example, the impact of real income on ecological footprint increases after the first income threshold level (θ=9.556, about US$14,129.22) is reached and the correlation is still positive. On the other hand, when the increase in real income reaches the second income threshold level (θ=10.865, about US$52,312.99), the impact of the real income increase on ecological footprint is weaker compared to the previous regime, but it is still positive. In conclusion, this shows that the ecological footprint does not decrease as the real income increases in the full sample. Therefore, this study shows that the EKC hypothesis is not valid in the full sample in terms of ecological footprint.

When countries are classified as advanced and non-advanced countries and the analysis is repeated, it is observed that the EKC hypothesis is not valid for advanced and non-advanced countries in terms of ecological footprint. The results reveal that increase in real income in advanced and non-advanced countries increases the ecological footprint, the impact of an increase in the real income on the ecological footprint gets weaker compared to the previous regime but remains positive after the first income threshold level ($\theta=10.137$, about US\$25,260.57 for advanced countries; $\theta=9.700$, about US\$16,317.61 for non-advanced countries) is reached in advanced countries while this impact increases compared to the previous regime and remains positive in non-advanced countries.

The results show that the ecological footprint does not decrease as the real income increases in the full sample as well as advanced and non-advanced countries and the EKC hypothesis is not valid. Global pollution can be one of the reasons why the EKC hypothesis is not valid (Chiu, 2017: 283). For example, even if a country tries to decrease the ecological footprint by reducing the CO_2 emission, air pollution can be affected by CO_2 emissions of neighboring countries and this can prevent the ecological footprint from decreasing. Helland and Whitford (2003) state that CO_2 emission in a border area in the USA is higher than CO_2 emission in an area that is not located in the border by 604 %. Another reason is that countries use energy as the main input in their production and the energy is produced by fossil fuels. Increased use of energy and thus fossil fuel affects economic growth of countries positively and cause ecological footprints to increase.

Conclusion

This study analyzes the validity of the EKC hypothesis for 80 countries in the period between 1975 and 2013 as well as the role of real income in the non-linear correlation between ecological footprint and real income. The Panel Smooth Transition Regression model developed by Gonzalez, Terasvirta and van Dijk (2005) was used in this study.

The findings of this study can be summarized as follows. Firstly, this study presents new evidence indicating that there is a non-linear correlation between ecological footprint and real income in the countries included in the study. Secondly, a positive correlation was determined between ecological footprint and real income. An increase in the real income initially increases the ecological footprint but after the real income reaches a certain level, the increase in the ecological footprint becomes slower. Therefore, this study shows that the EKC hypothesis is not valid in the analyzed countries in terms of ecological footprint. Thirdly, when we divide the sample into two groups as advanced and non-advanced countries, the direction of the correlation between ecological footprint and real income in advanced and non-advanced countries does not change and the EKC hypothesis is not valid in either group.

Appendix

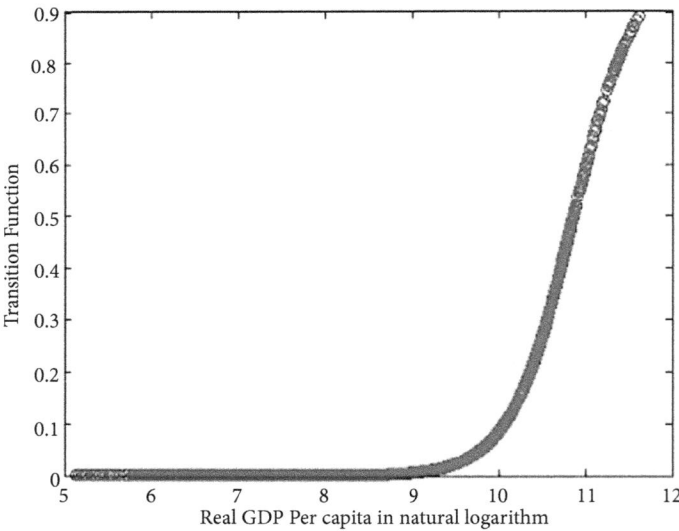

Fig. 1: Estimated Transition Function of the PSTR Model against Real GDP Per Capita for Full Sample. Source: Created by Authors

Economic Growth and Ecological Footprint 235

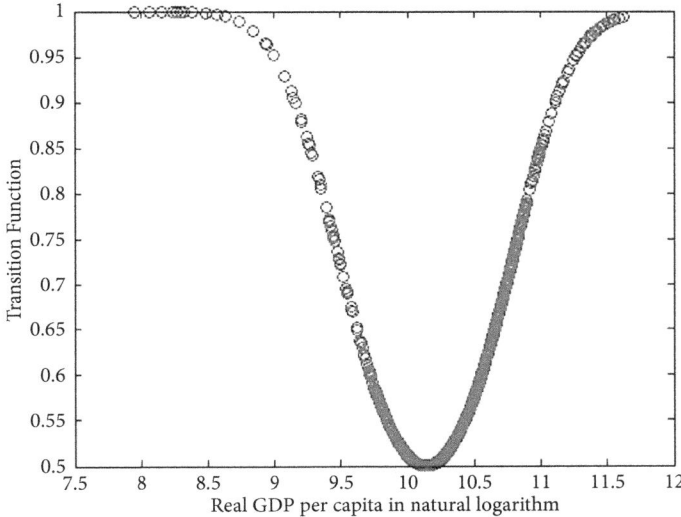

Fig. 2: Estimated Transition Function of the PSTR Model against Real GDP Per Capita for Advanced Countries. Source: Created by Authors

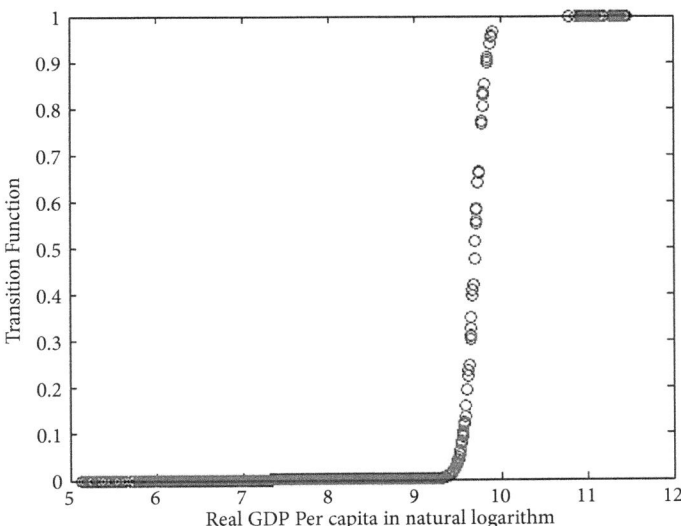

Fig. 3: Estimated Transition Function of the PSTR Model against Real GDP Per Capita for Non-advanced Countries. Source: Created by Authors

References

Agras, J. and Chapman, D. (1999) "A Dynamic Approach to The Environmental Kuznets Curve Hypothesis", Ecological Economics, 28(2): 267–277.

Albayrak, E. and Gökçe, A. (2015) "Ekonomik Büyüme ve Çevresel Kirlilik İlişkisi: Çevresel Kuznets Eğrisi ve Türkiye Örneği", Social Sciences Research Journal, 4(2): 279–301.

Al-Mulali, U., Weng-Wai, C., Sheau-Ting, L. ve Mohammed, A. H. (2015) "Investigating the Environmental Kuznets Curve (EKC) Hypothesis by Utilizing the Ecological Footprint as an Indicator of Environmental Degradation", Ecological Indicators, 48, 315–323.

Altıparmak, A. and Avcı, Z. (2011) "Uluslararası Ticaret, Ekolojik Ayakizi ve Türkiye", Ekonomi Bilimleri Dergisi", 3(2): 35–45.

Ang, J. B. (2007) "CO_2 Emissions, Energy Consumption, and Output in France", Energy Policy, 35(10): 4772–4778.

Aytun, C., Akın, C. S. and Algan, N. (2017) "Gelişen Ülkelerde Çevresel Bozulma, Gelir ve Enerji Tüketimi İlişkisi", Ömer Halisdemir Üniversitesi İktisadi ve İdari Bilimler Fakültesi Dergisi, 10(1): 1–11.

Bagliani, M., Bravo, G. ve Dalmazzone, S. (2008) "A Consumption-Based Approach to Environmental Kuznets Curves Using the Ecological Footprint Indicator", Ecological Economics, 65(3): 650–661.

Breusch, T. S. and Pagan, A. R. (1980). "The Lagrange Multiplier Test and Its Applications to Model Specification in Econometrics", The Review of Economic Studies, 47(1): 239–253.

Caviglia-Harris, J. L., Chambers, D. ve Kahn, J. R. (2009) "Taking the "U" Out of Kuznets: A Comprehensive Analysis of the EKC and Environmental Degradation", Ecological Economics, 68(4): 1149-1159.

Charfeddine, L. and Mrabet, Z. (2017) "The Impact of Economic Development and Social-Political Factors on Ecological Footprint: A Panel Data Analysis for 15 MENA Countries", Renewable and Sustainable Energy Reviews, 76, 138–154.

Chiu, Y. B. (2012) "Deforestation and the Environmental Kuznets Curve in Developing Countries: A Panel Smooth Transition Regression Approach", Canadian Journal of Agricultural Economics, 60(2): 177–194.

Chiu, Y. B. (2017) "Carbon Dioxide, Income and Energy: Evidence from A Non-Linear Model", Energy Economics, 61: 279–288.

Dasgupta, S., Benoıt, L., Wang, H. and Wheeler D. (2002) "Confronting the Environmental Kuznets Curve", Journal of Economic Perspectives, 16 (1): 147–168.

Dinda, S. (2004) "Environmental Kuznets Curve Hypothesis: A Survey", Ecological economics, 49(4): 431–455.

Erataş, F. and Uysal, D. (2014) "Çevresel Kuznets Eğrisi Yaklaşımının BRICT Ülkeleri Kapsamında Değerlendirilmesi", İktisat Fakültesi Mecmuası, 64(1): 1–25.

Ergün, S. and Polat Atay, M. (2015) "OECD Ülkelerinde Emisyonu, Elektrik Tüketimi ve Büyüme İlişkisi", Erciyes Üniversitesi İktisadi ve İdari Bilimler Fakültesi Dergisi, 45: 115–141.

Fouquau, J., Hurlin, C. and Rabaud, I. (2008), "The Feldstein–Horioka Puzzle: A Panel Smooth Transition Regression Approach", Economic Modelling, 25: 284–299.

Hansen, B. E. (1999). Threshold effects in non-dynamic panels: Estimation, testing, and inference. Journal of Econometrics, 93(2): 345-368.

Gill, A. R., Viswanathan, K. K. and Hassan, S. (2018) "A Test of Environmental Kuznets Curve (EKC) for Carbon Emission and Potential of Renewable Energy to Reduce Green Houses Gases (GHG) in Malaysia", Environment, Development and Sustainability, 20(3): 1103–1114.

Gonzalez, A., Terasvirta, T. and van Dijk, D. (2005), "Panel Smooth Transition Regression Model and an Application to Investment under Credit Constraint", Working Paper in Economics and Finance, Stockholm School of Economics, 64.

Grossman, G. M. ve Krueger, A. B. (1991) "Environmental Impacts of a North American Free Trade Agreement", NBER Working Papers, No: 3914.

Grossman, G. M. ve Krueger, A. B. (1995) "Economic Growth and the Environment", The Quarterly Journal of Economics, 110 (2): 353–377.

Gonzalez, A., Terasvirta, T. and Van Dijk, D. (2005) "Panel Smooth Transition Regression Models", Working Paper Series in Economics and Finance, No: 604.

Güriş, S. ve Tuna, E. (2011) "Çevresel Kuznets Eğrisinin Geçerliliğinin Panel Veri Modelleriyle Analizi", Trakya Üniversitesi Sosyal Bilimler Dergisi, 13(2): 173–190.

Güloğlu, B. ve Nazlioglu, Ş. (2013) "Impacts of Inflation on Agricultural Prices: Panel Smooth Transition Regression Analysis", Siyaset, Ekonomi ve Yönetim Araştırmaları Dergisi, 1(1): 1-20.

Harbaugh, W. T., Levinson, A. ve Wilson, D. M. (2002) "Reexamining the Empirical Evidence for an Environmental Kuznets Curve", The Review of Economics and Statistics, 84(3): 541–551.

Helland, E. ve Whitford, A. B. (2003) "Pollution Incidence and Political Jurisdiction: Evidence from the TRI", Journal of Environmental Economics and Management, 46(3): 403–424.

Herranz, A. A. ve Lorente, D. B. (2016) "Economic Growth and Energy Regulation in the Environmental Kuznets Curve", Environmental Science and Pollution Research, 23: 16478–16494.

Jalil A. ve Mahmud, S. F. (2009) "Environment Kuznets Curve for Emissions: A Cointegration Analysis for China", Energy Policy, 37: 5167–5172.

Karabıçak, M. and Armağan, R. (2004) "Çevre Sorunlarının Ortaya Çıkış Süreci, Çevre Yönetiminin Temelleri ve Ekonomik Etkileri", Süleyman Demirel Üniversitesi İİBF Dergisi, 9(2): 203–228.

Karaca, C. (2012) "Ekonomik Kalkınma ve Çevre Kirliliği İlişkisi: Gelişmekte Olan Ülkeler Üzerine Ampirik Bir Analiz", Ç.Ü. Sosyal Bilimler Enstitüsü Dergisi, 21(3): 139–156.

Kasman, A. ve Duman Y. (2015) "CO_2 Emissions, Economic Growth, Energy Consumption, Trade and Urbanization in New EU Member and Candidate Countries: A Panel Data Analysis", Economic Modelling, 44: 97–103.

Kearsley, A. ve Riddel, M. (2010) "A Further Inquiry into the Pollution Haven Hypothesis and the Environmental Kuznets Curve", Ecological Economics, 69(4): 905–919.

Lean, H. H. ve Smyth, R. (2010) "CO_2 Emissions, Electricity Consumption and Output in ASEAN", Applied Energy, 87(6): 1858–1864.

Li, T., Wang, Y. ve Zhao, D. (2016) "Environmental Kuznets Curve in China: New Evidence from Dynamic Panel Analysis", Energy Policy, 91: 138–147.

Mostafa, M. M. (2010) "A Bayesian Approach to Analyzing the Ecological Footprint of 140 Nations" Ecological Indicators, 10(4): 808–817.

Moon, H. R. ve Perron, B. (2004) "Testing for a unit root in panels with dynamic factors", Journal of Econometrics, 122(1): 81–126.

Moutinho, V., Varum, C. ve Madaleno, M. (2017) "How Economic Growth Affects Emissions? An Investigation of the Environmental Kuznets Curve in Portuguese and Spanish Economic Activity Sectors", Energy Policy, 106: 326–344.

Müller-Fürstenberger, G. ve Wagner, M. (2007) "Exploring the Environmental Kuznets Hypothesis: Theoretical and Econometric Problems", Ecological Economics, 62(3): 648–660.

Nasir, M. ve Rehman, F. U. (2011) "Environmental Kuznets Curve for Carbon Emissions in Pakistan: An Empirical Investigation", Energy Policy, 39(3): 1857–1864.

Özsoy, C. (2015) "Düşük Karbon Ekonomisi ve Türkiye'nin Karbon Ayakizi", Emek ve Toplum, 4(9): 199–215.

Panayotou, T. (1993) "Empirical Tests and Policy Analysis of Environmental Degradation at Different Stages of Economic Development", No: 238, Technology and Employment Programme, International Labour Organization, Geneva.

Pesaran, M. H., (2004), "General Diagnostic Tests for Cross Section Dependence in Panels", CEsifo Working Paper Series 1229, CEsifo Group Munich.

Pesaran, M. H., Ullah, A. ve Yamagata, T. (2008) "A bias- adjusted LM test of error cross- section independence", The Econometrics Journal. 11(1): 105–127.

Rees, W. E. ve Wackernagel, M. (1994) "Ecological Footprints and Appropriated Carrying Capacity: Measuring the Natural Capital Requirements of the Human Economy". In: Investing in Natural Capital: The Ecological Economics Approach to Sustainability, A.-M. Jansson, M. Hammer, C. Folke and R. Costanza (Eds.), 362–390, Washington, DC: Island Press.

Richmond, A. K. ve Kaufmann, R. K. (2006) "Is there a Turning Point in the Relationship Between Income and Energy Use and/or Carbon Emissions?", Ecological Economics, 56(2): 176–189.

Roca, J. (2003) "Do Individual Preferences Explain the Environmental Kuznets Curve?", Ecological Economics, 45(1): 3–10.

Sarısoy, S. ve Yıldız, F. (2013) "Karbondioksit (CO_2) Emisyonu ve Ekonomik Büyüme İlişkisi: Gelişmiş ve Gelişmekte Olan Ülkeler İçin Panel Veri Analizi", Namık Kemal Üniversitesi Sosyal Bilimler Metini, 2: 1–19.

Selden, T. M. ve Song, D. (1994) "Environmental Quality and Development: Is there A Kuznets Curve for Air Pollution Emissions?", Journal Of Environmental Economics and Management, 27(2): 147–162.

Shafik, N. ve Bandyopadhyay, S. (1992) "Economic Growth and Environmental Quality: Time-Series and Cross-Country Evidence", Policy Research Working Paper Series, No: 904, Washington, DC: World Bank Publications.

Shahbaz, M., Lean, H. H. ve Shabbir, M. S. (2012) "Environmental Kuznets Curve Hypothesis in Pakistan: Cointegration and Granger Causality", Renewable and Sustainable Energy Reviews, 16(5): 2947–2953.

Stern, D. I. (2004) "The Rise and Fall of the Environmental Kuznets Curve", World Development, 32(8): 1419–1439.

Stern, D. I. (2017) "The Environmental Kuznets Curve after 25 Years", Journal of Bioeconomics, 19(1): 7–28.

Vollebergh, H. R., Melenberg, B. ve Dijkgraaf, E. (2009) "Identifying Reduced-Form Relations with Panel Data: The Case of Pollution and Income", Journal of Environmental Economics and Management, 58(1): 27–42.

Wagner, M. (2008) "The Carbon Kuznets Curve: A Cloudy Picture Emitted by Bad Econometrics?", Resource and Energy Economics, 30(3): 388–408.

Wang, Y., Kang, L., Wu, X. ve Xiao, Y. (2013) "Estimating the Environmental Kuznets Curve for Ecological Footprint at the Global Level: A Spatial Econometric Approach", Ecological Indicators, 34, 15–21.

York, R., Rosa, E. A. ve Dietz, T. (2004) "The Ecological Footprint Intensity of National Economies", Journal of Industrial Ecology, 8(4): 139–154.

List of Figures

Fig. 1:	The Notion of Sustainable Marketing.	15
Fig. 1:	The Relationship between the Factors that Affect the Demand for Goods in General.	59
Fig. 1:	Industrial Revolutions One to Four.	104
Fig. 1:	The Common Interaction Area of Businesses.	121
Fig. 1:	Share of Energy Sources in Turkey in 2015.	151
Fig. 2:	GHG Emissions in Turkey.	161
Fig. 3:	GHGs Generated by Sectors, 2016.	162
Fig. 4:	Share of GHGs in Turkey in 2016.	163
Fig. 5:	CO_2 Emissions in Turkey.	163
Fig. 1:	Environmental Kuznets Curve.	189
Fig. 1:	Environmental Kuznets Curve.	205
Fig. 2:	Environmental Kuznets Curve (Low-Income Economies).	206
Fig. 3:	Environmental Kuznets Curve (Lower-Middle-Income Economies).	207
Fig. 4:	Environmental Kuznets Curve (Upper-Middle-Income Economies).	207
Fig. 1:	Estimated Transition Function of the PSTR Model against Real GDP Per Capita for Full Sample.	234
Fig. 2:	Estimated Transition Function of the PSTR Model against Real GDP Per Capita for Advanced Countries.	235
Fig. 3:	Estimated Transition Function of the PSTR Model against Real GDP Per Capita for Non-advanced Countries.	235

List of Tables

Tab. 1:	Sustainable Marketing Mix Elements.	16
Tab. 1:	Socio-Demographic Characteristics of the Participants.	60
Tab. 2:	The Statements Appear in the Scale and the Statistics of the Participants' Responses.	61
Tab. 3:	KMO and Bartlett Test of Sphericity.	62
Tab. 4:	Sub-Dimensions Obtained through Factor Analysis.	63
Tab. 5:	The Correlation between the Environmental Factors that Affect Consumers' Demand for Goods.	64
Tab. 6:	The Environmental Factors that Affect Consumers' Demand for Goods According to Age Groups.	65
Tab. 7:	The Environmental Factors that Affect Consumers' Demand for Goods According to Marital Status.	66
Tab. 8:	The Environmental Factors that Affect Consumers' Demand for Goods According to Consumers with Different Educational Backgrounds.	68
Tab. 9:	The Environmental Factors that Affect Consumers' Demand for Goods According to Different Income Levels.	69
Tab. 1:	Annual Greenhouse Gas Emissions by Sector, in 2010.	94
Tab. 2:	Components of Biogas.	95
Tab. 3:	Most Common Indicators of Sustainable Agriculture Themes among Existing Indices, Reports and Datasets Reviewed.	97
Tab. 1:	Comparison of the Conventional Management Mentality and Environmentally Conscious Management Mentality.	122
Tab. 2:	The Scope of the Environmental Accounting Concept. Source: EPA, 1995: 4	123
Tab. 3:	Obtaining Various Managerial Decision Benefits from Environmental Costs.	125
Tab. 4:	Classification of Private Costs.	129
Tab. 1:	Comparison of Annual Activity Report. Independent Environmental Report, Sustainability Report and Integrated Report.	139
Tab. 1:	GHGs and their Main Properties.	150
Tab. 2:	Summary from the Literature.	154
Tab. 1:	Used Variables and Expected Impact.	195
Tab. 2:	Statistics Summary Table.	195
Tab. 3:	Correlation Results.	195
Tab. 4:	Modified Walt Test Results.	195

Tab. 5:	Wooldridge Test.	196
Tab. 6:	Panel Data Estimation Table.	196
Tab. 7:	Fixed Effects.	198
Tab. 8:	Random Effects.	199
Tab. 1:	Variables and the Source of Dataset.	211
Tab. 2:	Descriptive Statistics.	212
Tab. 3:	Pairwise Correlations.	212
Tab. 4:	Cross Sectional Dependence and Coefficient Homogeneity Test Results.	214
Tab. 5:	Panel AMG Results.	215
Tab. 1:	Descriptive Statistics of Variables in Levels over the Period 1975–2013.	227
Tab. 2:	Cross Section Dependence.	228
Tab. 3:	Results of Moon and Perron's (2004) Panel Unit Root Tests.	229
Tab. 4:	Tests for the Linearity.	230
Tab. 5:	Tests for the Remaining Non-Linearity of the PSTR Model.	231
Tab. 6:	Estimated Results of the PSTR Model.	232

www.ingramcontent.com/pod-product-compliance
Ingram Content Group UK Ltd.
Pitfield, Milton Keynes, MK11 3LW, UK
UKHW021842210426
5322IPUK00022B/415